ELEMENTS OF
ENGINEERING STATICS

ELEMENTS OF ENGINEERING STATICS

by H. Deresiewicz
Associate Professor of Mechanical Engineering
Columbia University

NEW YORK
COLUMBIA UNIVERSITY PRESS

Copyright © 1957, 1958, 1959,
Columbia University Press, New York
First printing 1957
Second complete printing 1959

Manufactured in the United States of America

PREFACE

The present volume contains essentially the material discussed in the course on Engineering Statics given to pre-engineering students at Columbia University. It is assumed that the student has a working knowledge of elementary differential calculus and is concurrently studying integral calculus.

The first chapter deals with the algebra of vectors, considered at first without reference to subsequent applications. The purpose is to instill in the student a facility in manipulating vectors and thinking in terms of vectors. The reason is twofold: this approach will enhance understanding of some of the basic physical quantities with which the idea of direction is intrinsically associated, and it will simplify greatly the mathematical expression of such quantities and of the relations between them. This is particularly needed in the subsequent course which deals with kinematics and Newtonian dynamics of particles and rigid bodies.

Chapter II is devoted to consideration of the idea of a force and its moment, followed by a discussion of the concept of, and the conditions for, the equilibrium of a single particle and of systems of particles. The conditions arrived at are then applied, in Chapter III, to problems of equilibrium under concurrent, parallel, and general planar force systems, particular emphasis being placed on the idea of the "free-body" diagram.

Chapter IV offers a discussion of the concept of couples and of the equivalence of force systems and their reduction to simpler form. The techniques worked out in Chapter II and illustrated in Chapter III are applied, in Chapter V, to specific problems of statically determinate plane trusses and frames. Problems involving sliding friction are analyzed in Chapter VI. Chapter VII contains a definition of the concept of work done by a force, which leads, in turn, to a discussion of the principle of virtual work and its application to problems of static equilibrium. Moreover, in this chapter the idea of a conservative system and its potential energy is introduced and then employed in the analysis of the stability of the equilibrium of such a system. Chapter VIII has a brief exposition of simple spatial systems. The volume concludes with an Appendix which deals with the concept of the center of gravity and its application to problems of distributed loading.

Throughout the text theoretical concepts are illustrated by specific examples in accordance with the primary purpose of the course, which is to foster understanding of the fundamental principles of statics and to develop a facility in solving problems arising in engineering practice.

I wish to express my gratitude to Mr. C. W. Thurston, of the Department of Civil Engineering, for his valuable suggestions. To Professor J. E. Englund, of the Department of Mechanical Engineering, my warmest thanks are due for his encouragement and support.

Columbia University
New York, New York
July, 1957

H. Deresiewicz

CONTENTS

I. ELEMENTS OF VECTOR ALGEBRA 1
  1. Introduction 1
  2. Equality of Vectors 2
  3. Addition of Vectors 2
  4. Subtraction of Vectors 3
  5. Multiplication by Scalars 4
  6. Parallel Projection of Vectors 4
  7. Unit Coordinate Vectors 5
  8. Scalar Multiplication of Vectors 8
  9. Vector Multiplication of Two Vectors 10
  10. Moment of a Vector 12
  11. Differentiation of Vectors 17

II. THE PROBLEM OF EQUILIBRIUM 21
  1. Introduction 21
  2. Composition of Forces and Moments of Forces 21
  3. Newton's Laws for a Particle 27
  4. Equilibrium 28

III. EQUILIBRIUM OF SIMPLE PLANAR SYSTEMS 32
  1. Introduction 32
  2. Free-Body Diagram 32
  3. Concurrent Forces 34
  4. Parallel Forces 37
  5. General Coplanar Forces 38
  6. Static Determinateness and Indeterminateness 39

IV. EQUIVALENCE OF FORCE SYSTEMS 41
  1. Introduction 41
  2. Couples 42
  3. Reduction of a System of Forces 43

CONTENTS

V. SIMPLE STRUCTURES	46
1. Trusses	46
2. The Method of Joints	48
3. The Method of Sections	51
4. Complex Trusses	52
5. Frames	54
VI. SLIDING FRICTION	58
1. Introduction	58
2. The Laws of Friction	58
3. Applications of Coulomb's Laws of Friction	60
VII. WORK AND ENERGY METHODS	65
1. Work	65
2. The Principle of Virtual Work (for a Free Particle)	70
3. Ideal Constraints	70
4. The Principle of Virtual Work (for Systems)	73
5. Potential Energy	79
6. Stability	81
VIII. EQUILIBRIUM OF SIMPLE SPATIAL SYSTEMS	86
Appendix. CENTER OF GRAVITY	90
1. Static Equivalence of Parallel Forces	90
2. Center of Gravity	91
3. Application to Problems Involving Distributed Load	94
PROBLEMS	97

Chapter I. ELEMENTS OF VECTOR ALGEBRA

1. Introduction

The quantities with which we shall be concerned may be classified into *vectors* and *scalars*, depending on whether or not the idea of direction is associated with them.

A quantity which is completely specified by a single number, positive or negative, without intrinsic reference to direction in space, is called a *scalar*. Such quantities as length and time belong to this category.

A *vector* quantity, on the other hand, involves inherently the idea of direction as well as magnitude. For example, the position of a point B relative to another point A may be specified by means of a straight line drawn from A to B (Fig. 1). Moreover, it may equally well be specified by any equal and parallel straight line drawn in the same sense from,

Fig. 1

say, C to D, since the position of D relative to C is the same as that of B relative to A. A straight line regarded in this way as having a definite magnitude and direction, but no specific location in space, is called a *vector* —occasionally, to emphasize the latter property, a *free vector*. In short, a vector may be represented geometrically by a directed line segment. For example, "translation" of a body, accomplished by a displacement in which lines joining initial and final positions of various points of the body are all equal and parallel, is specified by a free vector which may

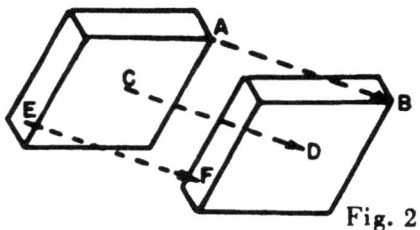

Fig. 2

be any one of these lines (Fig. 2).

In distinction, a *sliding* vector is one whose point of application is confined to any point on a given line; a *bound* vector is one with a unique point of application. In general, by the term *vector*, we will understand a free vector.

Notation: in print, vectors are denoted by boldface letters (e.g., **P**), scalars by italicized letters (e.g., m); in manuscript, a variety of notations is employed for vectors, as for example letters with superscribed arrows (e.g., $\vec{P}$) or letters with a line doubled (e.g., ⱣⱣ). It will be well for the student to achieve a facility in writing vectors; this may be accomplished most readily by intensive practice. For convenience, sample Latin alphabets are given in capital and lower case letters.

A B C D E F G H I J K L M
N O P Q R S T U V W X Y Z
a b c d e f g h i j k l m
n o p q r s t u v w x y z

2. Equality of Vectors

Two vectors **P** and **Q** are regarded as equal, or rather identical, if and only if they have the same magnitude and direction. We use the equation **P** = **Q** to express this complete identity. Here is the definition of the equal sign (=) as used in the present connection; note that there can be no question of equality between vectors whose directions are different.

Straight lines which are equal and parallel to the same straight line are equal and parallel to one another. It follows that if **P** = **R** and **Q** = **R**, then **P** = **Q**; that is, vectors which are equal to the same vector are equal to one another.

3. Addition of Vectors

Consider the composition of displacements of a point located initially at A (Fig. 3). For example, if such a point is displaced from A to B and then from B to C, the final result is equivalent to a displacement represented by AC. Hence, we speak of AC as the *sum* (in the sense of *geometric sum*) of vectors AB and BC, and write $AB + BC = AC$.

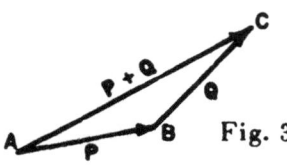

Fig. 3

To construct the sum of any two given vectors **P** and **Q**, draw line AB to represent **P**, then BC to represent **Q**; the sum **P** + **Q** is then represented by AC.

Let us see if this definition of vector addition follows the rules which govern ordinary algebraic addition. If, in Fig. 3, we complete the parallelogram $ABCD$ (Fig. 4), we have

$$DC = AB = \mathbf{P}$$
$$AD = BC = \mathbf{Q}$$

Therefore, **Q** + **P** = $AD + DC = AC$, or

$$\mathbf{Q} + \mathbf{P} = \mathbf{P} + \mathbf{Q}$$

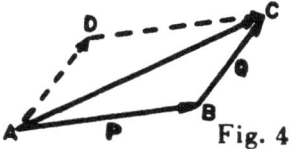

Fig. 4

This is the *commutative* law of addition for vectors. The result is not self-evident, but is a consequence of Euclid's theory of parallels.

When we wish to indicate that a particular vector arises as the sum of two vectors **P** and **Q**, we enclose the sum in parentheses, as (**P** + **Q**). Accordingly, there is a distinction in the meaning of (**P** + **Q**) + **R** and **P** + (**Q** + **R**). Thus, if we set (Fig. 5) $AB = \mathbf{P}$, $BC = \mathbf{Q}$, $CD = \mathbf{R}$ we have

$$(\mathbf{P} + \mathbf{Q}) + \mathbf{R} = AC + CD = AD$$
$$\mathbf{P} + (\mathbf{Q} + \mathbf{R}) = AB + BD = AD$$

Hence,

$$(\mathbf{P} + \mathbf{Q}) + \mathbf{R} = \mathbf{P} + (\mathbf{Q} + \mathbf{R})$$

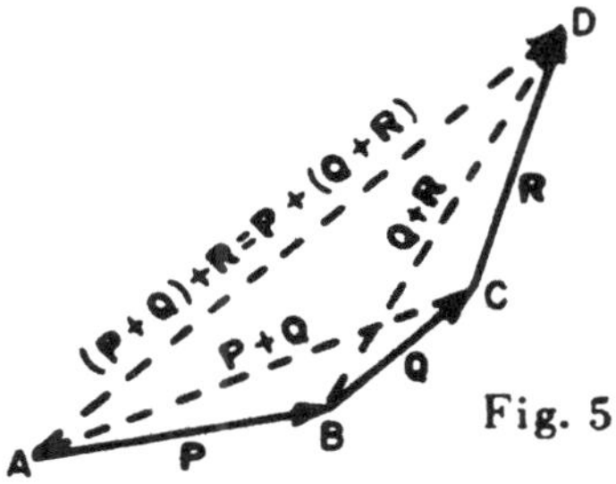

This is the *associative* law of addition for vectors. It follows from this and the commutative law that three or more vectors may be added in any order without affecting the result. Hence, in practice, the parentheses, which are, strictly speaking, necessary to define the succession of operations, are omitted, e.g.,

$$(\mathbf{P} + \mathbf{Q}) + (\mathbf{R} + \mathbf{S}) = \mathbf{P} + \mathbf{R} + \mathbf{Q} + \mathbf{S} = \mathbf{Q} + \mathbf{S} + \mathbf{P} + \mathbf{R}, \text{ etc.}$$

Fig. 5

It should be noted that, in Fig. 5, points A, B, C, D need not be coplanar (i.e., in the same plane); the vectors **P**, **Q**, **R** may have any direction whatever in space.

4. Subtraction of Vectors

The *minus* sign (−) prefixed to a vector is employed to indicate that the direction of the vector is reversed; for example, $BA = -AB$. For brevity, we write **P** − **Q** for **P** + (−**Q**). Hence, in the parallelogram in Fig. 4, $\mathbf{P} - \mathbf{Q} = AB - BC = AB + CB = DA + AB = DB$. (If **P** and **Q** denote translations of two bodies, the vector **P** − **Q** represents the displacement of the first body relative to the second. The simplest example of this is the case of one-dimensional translation, i.e., one in which the displacements occur along the same straight line.)

A vector whose terminal points coincide is called a *null vector* and is denoted by the symbol 0. All such vectors may, clearly, be regarded as equivalent. Thus, in the parallelogram in Fig. 4, $AB + BC + CA = 0$, $AB + BA = 0$, $AA = 0$. Moreover, $AB + BB = AB$, or $\mathbf{P} + 0 = \mathbf{P}$. Hence, also (**P** − **Q**) + **Q** = **P** + (−**Q** + **Q**) = **P** + 0 = **P**. This is the fundamental property of the minus sign in formal algebra.

5. Multiplication by Scalars

We define the product of the scalar m and the vector **P**, denoted by m**P**, to be a vector having the following properties: its magnitude is given by the product of P (the magnitude of **P**) and $|m|$ (the absolute value of m, i.e., its value without regard to algebraic sign); its direction is that of **P** if m is positive, or the reverse if m is negative. This is illustrated in

Fig. 6

Fig. 6, in which $AB =$ **P**, $AC =$ 2**P** (i.e., $m = +2$) and $AD = -\frac{3}{2}$ **P** (i.e., $m = -\frac{3}{2}$).

It follows from the definition that if **P** = **Q** then m**P** = m**Q**; further, m**P** + n**P** = $(m + n)$**P**.

It remains for us to examine whether the distributive law, $m(\mathbf{P} + \mathbf{Q}) = m\mathbf{P} + m\mathbf{Q}$, which is fundamental in ordinary algebra, holds for vectors. In Fig. 7, let $OA =$ **P**, $OA' = m\mathbf{P}$, $AB =$ **Q**, $A'B' = m\mathbf{Q}$.

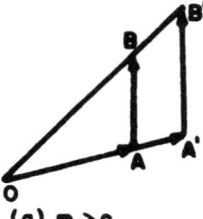 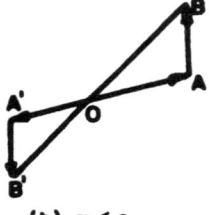

(a) m > 0 (b) m < 0 Fig. 7

By construction,

$$\frac{OA'}{OA} = m = \frac{A'B'}{AB}, \quad \frac{OA'}{A'B'} = \frac{OA}{AB}.$$

Further, since $AB \parallel A'B'$, $\angle OA'B' = \angle OAB$. Therefore, $\triangle OAB \sim \triangle OA'B'$ and, hence, points O, B, B' are collinear (as drawn). It follows that

$$\frac{OB'}{OB} = \frac{OA'}{OA} = m,$$

so that $OA' + A'B' = OB' = mOB$, or

$$m\mathbf{P} + m\mathbf{Q} = m(\mathbf{P} + \mathbf{Q}).$$

6. Parallel Projection of Vectors

The projection of a point A on a given straight line OX is defined as the point A' at which a line (in three dimensions, a plane) through A in some prescribed direction (α) meets OX (Fig. 8). If AB represents a given vector and A' and B' are the projections of the points A and B on line OX, the vector $A'B'$ is called the *vector component* of AB on OX and the magnitude of $A'B'$ is called the *scalar component* of AB on OX.

ELEMENTS OF VECTOR ALGEBRA

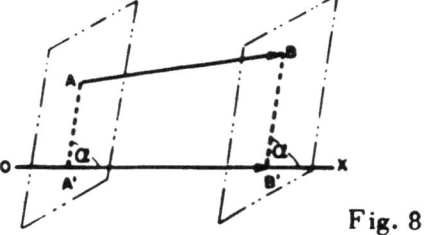

Fig. 8

Theorem: The projection of the sum of two or more vectors is equal to the sum of the projections of the several vectors.

Thus, if AB, BC are drawn to represent any two vectors (Fig. 9), and A', B', C' are the projections on OX of A, B, C, respectively, then $A'B' + B'C = A'C'$. But $A'C'$ is the projection on OX of the vector AC, which is the sum of the vectors AB and BC.

The case of greatest importance is that of *orthogonal* projection, in which the projecting planes (or lines) are perpendicular to OX (i.e., $\alpha = 90°$). We confine our attention exclusively to orthogonal projections.

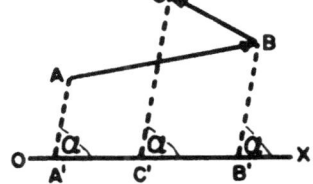

Fig. 9

7. Unit Coordinate Vectors

Any vector whose magnitude, in some consistent set of units, is unity is called a *unit* vector. It should be noted that two unit vectors are, in general, not equal, since their directions will not be the same.

A unit vector in the direction of a vector **P** will be denoted by $\mathbf{P_0}$. Hence, any vector **P** may be written in the form $\mathbf{P} = P\mathbf{P_0}$.

Consider a given line OX and fix one direction, say that from O to X, as positive; hence, the opposite direction is negative. We denote the unit vector in the positive direction along OX by **i**. In Fig. 10, the vector **Q′** represents the projection of a given vector **P** on OX. The magnitude of vector **Q′** is given by the length $A'B'$, or $P \cos \theta$, θ being the angle which **P** makes with OX. Hence, $\mathbf{Q'} = (P \cos \theta)\mathbf{i}$.

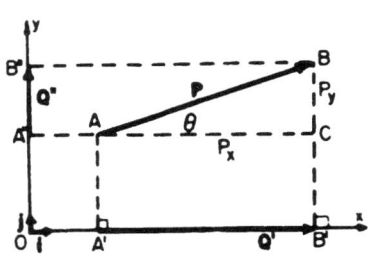

Fig. 10

We see that the scalar component of **P** on OX (denoted henceforth by P_x) is given by the product of the magnitude of **P** and the cosine of the angle between **P** and **i**.

Now assume Fig. 10 to lie in the plane of the paper and erect $OY \perp OX$; the unit vector in the positive direction (from O to Y) is denoted by **j**. Then the scalar component of **P** on OY (length $A''B''$, denoted by P_y) is given by the product of the magnitude of **P** and the cosine of the angle between **P** and **j**, i.e., $P \cos(90 - \theta) = P \sin \theta$, and the corresponding vector component is $\mathbf{Q}'' = (P \sin \theta)\mathbf{j}$.

We may express the (planar) vector **P** in terms of the unit vectors **i** and **j** by noting that

$$\mathbf{P} = \mathbf{Q}' + \mathbf{Q}'' = P_x \mathbf{i} + P_y \mathbf{j}$$

where $P_x = P \cos \theta$, $P_y = P \sin \theta$.

From rt. $\triangle ACB$, the magnitude and direction of **P** may be expressed in terms of its scalar components P_x and P_y; thus, the magnitude is

$$P = \sqrt{P_x^2 + P_y^2}$$

and the direction is given by

$$\theta = \text{arc tan } \frac{P_y}{P_x}.$$

Let us generalize these results to three-dimensional space (Fig. 11). The unit vectors along OX, OY, OZ are denoted by **i**, **j**, **k** respectively. A vector **P** (OA) makes, respectively, angles α, β and γ with the coordinate axes x, y, z, as shown. Its scalar components on these axes are denoted by $P_x (OB)$, $P_y (OC \text{ or } BA')$ and $P_z (OD \text{ or } A'A)$. Hence, we may express **P** in terms of its components as follows:

$$OA = OB + BA' + A'A$$
$$\text{or} \quad \mathbf{P} = P_x \mathbf{i} + P_y \mathbf{j} + P_z \mathbf{k}$$

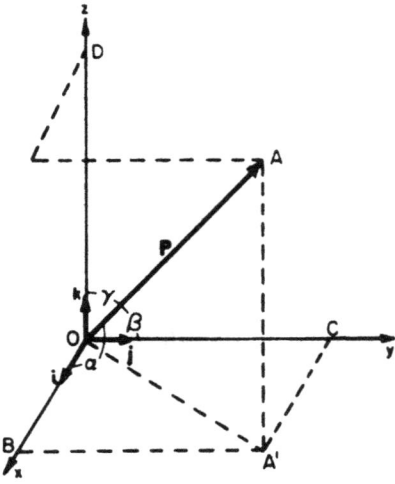

Fig. 11

where $P_x = P \cos \alpha$, $P_y = P \cos \beta$, $P_z = P \cos \gamma$.

To express the magnitude of **P**, we note that, in rt. $\triangle OA'A$, $(OA)^2 = (OA')^2 + (A'A)^2$. But, from rt. $\triangle OBA'$, $(OA')^2 = (OB)^2 + (BA')^2$. Hence, $(OA)^2 = (OB)^2 + (BA')^2 + (A'A)^2$, or $P^2 = P_x^2 + P_y^2 + P_z^2$. Inserting the values of P_x, P_y, and P_z in terms of α, β, γ, and introducing the customary symbols for direction cosines, viz., $l \equiv \cos \alpha$, $m \equiv \cos \beta$, $n \equiv \cos \gamma$, we obtain

$$\cos^2 \alpha + \cos^2 \beta + \cos^2 \gamma = l^2 + m^2 + n^2 = 1.$$

This is the well-known result from analytic geometry which states that the sum of the squares of the direction cosines of any line must equal unity.

To summarize, given the magnitude (P) and direction cosines (l, m, n) of a vector, we may find its scalar components on a set of orthogonal axes from the relations

$$P_x = Pl, \ P_y = Pm, \ P_z = Pn.$$

Conversely, given the scalar components (P_x, P_y, P_z) of a vector, its magnitude and direction may be found from

$$P = \sqrt{P_x^2 + P_y^2 + P_z^2},$$

$$l = \frac{P_x}{P}, \quad m = \frac{P_y}{P}, \quad n = \frac{P_z}{P}.$$

Example 1: It is known of a certain vector that its magnitude is $P = 100$, the angles it forms with the positive X and Y axes are, respectively, $\alpha = 30°$, $\beta = 70°$, and it is located in the first octant. It is required to express the vector in terms of the unit coordinate vectors **i**, **j**, **k**.

Given: $P = 100$, $\alpha = 30°$, $\beta = 70°$, vector in first octant.
Hence, $\qquad l \equiv \cos \alpha = 0.866, \quad m \equiv \cos \beta = 0.342$
Further, $\qquad\qquad l^2 + m^2 + n^2 = 1,$
or $\qquad\qquad n^2 = 1 - l^2 - m^2 = 0.133$
so that $\qquad\qquad n \equiv \cos \gamma = +0.365$
Therefore, $\qquad P_x = Pl = 86.6$
$\qquad\qquad\qquad P_y = Pm = 34.2$
$\qquad\qquad\qquad P_z = Pn = 36.5$
and $\qquad\qquad$ **P** $= 86.6$**i** $+ 34.2$**j** $+ 36.5$**k**.

To find the expression for a unit vector along **P**, we recall that **P** $= P_x$**i** $+ P_y$**j** $+ P_z$**k** and **P** $= P$**P**$_0$. Hence,

$$\mathbf{P}_0 = \frac{\mathbf{P}}{P} = \frac{P_x \mathbf{i} + P_y \mathbf{j} + P_z \mathbf{k}}{P}$$

or
$$\mathbf{P}_0 = l\mathbf{i} + m\mathbf{j} + n\mathbf{k}.$$

Example 2: Given a vector whose **i**, **j**, **k** components are of magnitude 5, 0, 12, respectively; find the magnitude and direction of this vector and the unit vector parallel to it.

Given: $P_x = 5$, $P_y = 0$, $P_z = 12$.

Hence,
$$P = \sqrt{P_x^2 + P_y^2 + P_z^2} = 13.$$

$$l = \frac{P_x}{P} = \frac{5}{13} = 0.3846; \quad \therefore \alpha = 67°23'.$$

$$m = \frac{P_y}{P} = 0; \quad \therefore \beta = 90°.$$

$$n = \frac{P_z}{P} = \frac{12}{13} = 0.9231; \quad \therefore \gamma = 22°37'.$$

The corresponding unit vector is
$$\mathbf{P}_0 = l\mathbf{i} + m\mathbf{j} + n\mathbf{k} = \tfrac{1}{13}(5\mathbf{i} + 12\mathbf{k}).$$

It is easy to see that, if
$$\mathbf{P} = P_x\mathbf{i} + P_y\mathbf{j} + P_z\mathbf{k},$$
$$\mathbf{Q} = Q_x\mathbf{i} + Q_y\mathbf{j} + Q_z\mathbf{k},$$
$$\mathbf{R} = R_x\mathbf{i} + R_y\mathbf{j} + R_z\mathbf{k},$$
then
$$\mathbf{P} \pm \mathbf{Q} \pm \mathbf{R} \pm \ldots = (P_x \pm Q_x \pm R_x \pm \ldots)\mathbf{i}$$
$$+ (P_y \pm Q_y \pm R_y \pm \ldots)\mathbf{j}$$
$$+ (P_z \pm Q_z \pm R_z \pm \ldots)\mathbf{k}.$$

The result may be stated in the form of the

Theorem: the components of the sum (difference) of a number of vectors are given by the sum (difference) of the respective components of the vectors.

We have seen (Section 2) that equality of two vectors presupposes equality of their respective magnitudes and directions. This, in turn, implies equality of their respective scalar components along any set of orthogonal coordinate axes. In the special case of a null vector, each scalar component is identically zero.

8. Scalar Multiplication of Vectors

Given two vectors, **a** and **b**, their *scalar* (or *dot*) product is defined by
$$\mathbf{a} \cdot \mathbf{b} = ab\cos(\mathbf{a}, \mathbf{b}).$$

ELEMENTS OF VECTOR ALGEBRA 9

Expressed in words, the dot product of two vectors is a scalar defined as the product of the magnitudes of the two vectors and of the cosine of the angle between them.

The following theorems derive from the definition of the dot product:

(1) $\mathbf{a}\cdot\mathbf{b} = \mathbf{b}\cdot\mathbf{a}$. The operation is commutative.

(2) $\mathbf{a}\cdot\mathbf{a} = a^2$.

(3) For unit vectors, $\mathbf{a}_0\cdot\mathbf{b}_0 = \cos(\mathbf{a},\mathbf{b})$.

In particular, for Cartesian unit vectors,

$$\mathbf{i}\cdot\mathbf{i} = \mathbf{j}\cdot\mathbf{j} = \mathbf{k}\cdot\mathbf{k} = 1$$
$$\mathbf{i}\cdot\mathbf{j} = \mathbf{j}\cdot\mathbf{k} = \mathbf{k}\cdot\mathbf{i} = 0.$$

(4) If $\mathbf{a}\cdot\mathbf{b} = 0$ and $\mathbf{a}, \mathbf{b} \neq 0$, then $\mathbf{a} \perp \mathbf{b}$. Conversely, if $\mathbf{a} \perp \mathbf{b}$, then $\mathbf{a}\cdot\mathbf{b} = 0$.

(5) $\mathbf{a}\cdot(\mathbf{b}+\mathbf{c}) = \mathbf{a}\cdot\mathbf{b} + \mathbf{a}\cdot\mathbf{c}$. The operation is distributive.

Corrolary: $(\mathbf{a}+\mathbf{b})\cdot(\mathbf{c}+\mathbf{d}) = \mathbf{a}\cdot\mathbf{c} + \mathbf{a}\cdot\mathbf{d} + \mathbf{b}\cdot\mathbf{c} + \mathbf{b}\cdot\mathbf{d}$.

(6) Any vector can be expressed in the form

$$\mathbf{P} = (\mathbf{P}\cdot\mathbf{i})\mathbf{i} + (\mathbf{P}\cdot\mathbf{j})\mathbf{j} + (\mathbf{P}\cdot\mathbf{k})\mathbf{k}.$$

Proof: Let $\mathbf{P} = P_x\mathbf{i} + P_y\mathbf{j} + P_z\mathbf{k}$; then

$$\mathbf{P}\cdot\mathbf{i} = P_x, \quad \mathbf{P}\cdot\mathbf{j} = P_y, \quad \mathbf{P}\cdot\mathbf{k} = P_z,$$

which demonstrates the theorem.

In general, the scalar component (i.e., the magnitude of the projection) of a vector $\mathbf{P}$ on a line whose direction is denoted by the unit vector $\mathbf{a}_0$ is $\mathbf{P}\cdot\mathbf{a}_0$. This follows immediately from the definition of the dot product and the meaning of a scalar component on a line.

Given vectors $\mathbf{a}$ and $\mathbf{b}$, expressed in component form, we wish to derive the form their dot product will take. Thus, suppose

$$\mathbf{a} = a_x\mathbf{i} + a_y\mathbf{j} + a_z\mathbf{k}$$
and
$$\mathbf{b} = b_x\mathbf{i} + b_y\mathbf{j} + b_z\mathbf{k};$$

we find, upon applying the distributive law—Theorem (5) above—,

$$\mathbf{a}\cdot\mathbf{b} = a_x b_x + a_y b_y + a_z b_z.$$

Hence the dot product of two vectors is given by the sum of the products of their corresponding components.

From the definition of dot product, we find

$$\cos(\mathbf{a},\mathbf{b}) = \frac{\mathbf{a}\cdot\mathbf{b}}{ab} = \frac{a_x b_x + a_y b_y + a_z b_z}{ab}$$
$$= l_a l_b + m_a m_b + n_a n_b.$$

This is the well-known formula from analytic geometry for the angle between two lines in terms of their direction cosines. In this connection, and for later use, we recall that a line in space may be specified by two points through which it passes. Let the coordinates of such points A_1 and A_2 be (x_1, y_1, z_1) and (x_2, y_2, z_2), respectively (Fig. 12). Then the direction cosines are given by

$$l = \frac{x_2 - x_1}{d}, \quad m = \frac{y_2 - y_1}{d}, \quad n = \frac{z_2 - z_1}{d},$$

where d represents the length $A_1 A_2$, or

$$d = \sqrt{(x_2 - x_1)^2 + (y_2 - y_1)^2 + (z_2 - z_1)^2}.$$

The vector $A_1 A_2$, whose scalar components along the x, y, z axes are ld, md, nd, respectively, may, accordingly, be expressed by

$$A_1 A_2 = (x_2 - x_1)\mathbf{i} + (y_2 - y_1)\mathbf{j} + (z_2 - z_1)\mathbf{k}.$$

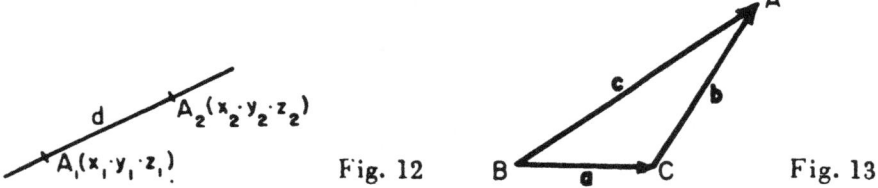

Fig. 12   Fig. 13

As another illustration of the application of scalar multiplication to geometry, we shall prove the law of cosines. Referring to Fig. 13, we see that $\mathbf{c} = \mathbf{a} + \mathbf{b}$; then,

$$\begin{aligned}\mathbf{c} \cdot \mathbf{c} = c^2 &= \mathbf{a} \cdot \mathbf{a} + \mathbf{a} \cdot \mathbf{b} + \mathbf{b} \cdot \mathbf{a} + \mathbf{b} \cdot \mathbf{b} \\ &= a^2 + b^2 + 2ab \cos(\mathbf{a}, \mathbf{b}) \\ &= a^2 + b^2 - 2ab \cos C.\end{aligned}$$

Example: Show that the vectors

$$\mathbf{a} = (T \cos \theta)\mathbf{i} + (T \sin \theta)\mathbf{j}$$
$$\mathbf{b} = (S \sin \theta)\mathbf{i} - (S \cos \theta)\mathbf{j}$$

are perpendicular.

The scalar product $\mathbf{a} \cdot \mathbf{b} = (T \cos \theta)(S \sin \theta) + (T \sin \theta)(-S \cos \theta) = 0$. Since neither $\mathbf{a}$ nor $\mathbf{b}$ is a null vector, it follows, by Theorem (4) of this section, that $\mathbf{a} \perp \mathbf{b}$.

9. Vector Multiplication of Two Vectors

Given two vectors, their *vector* (or *cross*) product is defined by

ELEMENTS OF VECTOR ALGEBRA

$\mathbf{a} \times \mathbf{b}$ = a vector whose $\begin{cases} \text{magnitude} = ab \sin{(\mathbf{a},\mathbf{b})}; \\ \text{direction} = \perp \text{plane containing } \mathbf{a} \text{ and } \mathbf{b}; \\ \text{sense} = \text{sense of advance of a righthand screw,} \\ \qquad\quad \text{rotated from } \mathbf{a} \text{ to } \mathbf{b} \text{ through the angle} \\ \qquad\quad \text{smaller than } 180°. \end{cases}$

The spatial relations inherent in this definition are illustrated in Fig. 14.

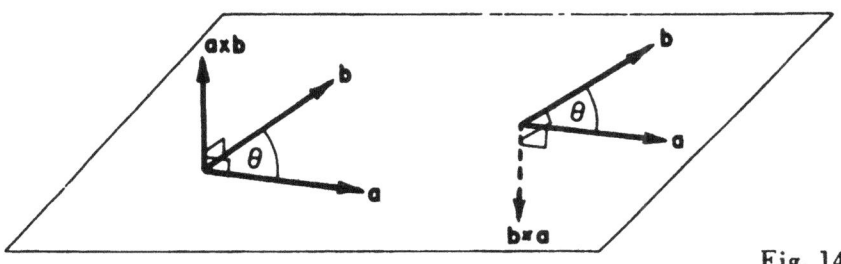

Fig. 14

The following theorems derive from the definition of a cross product:

(1) $\mathbf{a} \times \mathbf{b} = -\mathbf{b} \times \mathbf{a}$. The cross product is *non-commutative*.
(2) $\mathbf{a} \times \mathbf{a} = \mathbf{a} \times (-\mathbf{a}) = 0$.

In particular, $\mathbf{i} \times \mathbf{i} = \mathbf{j} \times \mathbf{j} = \mathbf{k} \times \mathbf{k} = 0$

(3) $\left.\begin{array}{l} \mathbf{i} \times \mathbf{j} = \mathbf{k} \\ \mathbf{j} \times \mathbf{k} = \mathbf{i} \\ \mathbf{k} \times \mathbf{i} = \mathbf{j} \end{array}\right\} \begin{array}{l} \text{cyclical} \\ \text{order} \\ (x,y,z) \end{array}$ $\left.\begin{array}{l} \mathbf{i} \times \mathbf{k} = -\mathbf{j} \\ \mathbf{j} \times \mathbf{i} = -\mathbf{k} \\ \mathbf{k} \times \mathbf{j} = -\mathbf{i} \end{array}\right\} \begin{array}{l} \text{non-cyclical} \\ \text{order} \end{array}$

(4) If $\mathbf{a} \times \mathbf{b} = 0$ and $\mathbf{a}, \mathbf{b} \neq 0$, then $\mathbf{a} \parallel \mathbf{b}$. Conversely, if $\mathbf{a} \parallel \mathbf{b}$, then $\mathbf{a} \times \mathbf{b} = 0$.

(5) $\mathbf{a} \times (\mathbf{b} + \mathbf{c}) = \mathbf{a} \times \mathbf{b} + \mathbf{a} \times \mathbf{c}$. The operation is distributive, but the order of factors must be preserved.

In rectangular component form,

$$\mathbf{a} \times \mathbf{b} = (a_x \mathbf{i} + a_y \mathbf{j} + a_z \mathbf{k}) \times (b_x \mathbf{i} + b_y \mathbf{j} + b_z \mathbf{k})$$
$$= (a_y b_z - a_z b_y)\mathbf{i} + (a_z b_x - a_x b_z)\mathbf{j} + (a_x b_y - a_y b_x)\mathbf{k}.$$

Given a parallelogram whose sides have lengths a and b and an included angle θ (Fig. 15), its area is given by $ab \sin \theta$. Alternatively, this is represented by the magnitude of the vector $\mathbf{a} \times \mathbf{b}$.

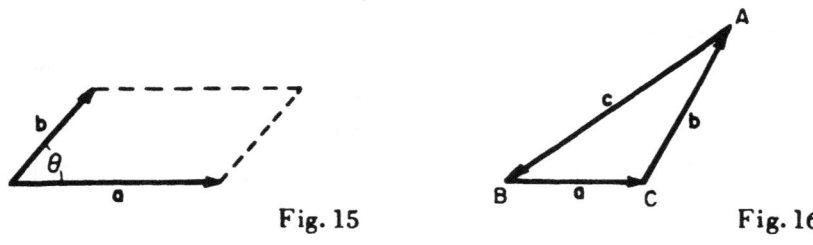

Fig. 15   Fig. 16

As an illustration of the application of the cross product, we shall derive the law of sines. Referring to Fig. 16, we note $a + b + c = 0$; hence, operating by $a \times$, we obtain

$$a \times b + a \times c = 0$$

or $$[ab \sin (180 - C)] n_0 - [ac \sin (180 - B)] n_0 = 0,$$

where n_0 represents a unit vector directed out of the plane of the paper. Therefore,

$$(b \sin C - c \sin B) n_0 = 0,$$

and, since $n_0 \neq 0$,

$$\frac{b}{\sin B} = \frac{c}{\sin C}.$$

10. Moment of a Vector

Consider a point A whose rectangular coordinates with respect to an arbitrary origin O are (x, y, z). The vector OA is called the *position vector* of A with respect to O (Fig. 17) and is expressed by

$$r = xi + yj + zk.$$

Now, let P be a vector passing through A; then the *moment of P about point O* is a vector defined by

$$M = r \times P.$$

Thus, the magnitude of M is $rP \sin (r, P)$ and the direction is perpendicular to the plane of r and P, in the sense of advance of a righthand screw rotated from r to P (Fig. 17). It may be noted that the magnitude of the moment M is equal to the area of the parallelogram $OABC$.

We recall (see Section 8) that the magnitude of the projection of a vector on a line is given by the dot product of the vector and a unit vector along the line. Hence, the projection of the moment M on, say, the

ELEMENTS OF VECTOR ALGEBRA

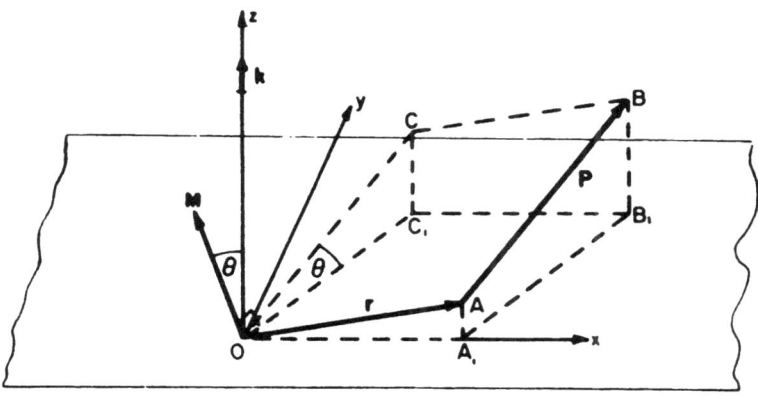

Fig. 17

z-axis is given by the scalar product $\mathbf{M} \cdot \mathbf{k} = M \cos \theta$. Geometrically, this is represented by the area of the projected parallelogram $OA_1B_1C_1$. In general, the projection of the moment $\mathbf{M}$ on any line L, passing through O, is given by $\mathbf{M} \cdot \mathbf{L_0}$, where $\mathbf{L_0}$ is a unit vector along L. This quantity is called the *(scalar) moment of vector* $\mathbf{P}$ *about line* L.

Example 1: Fig. 18 represents a cube of side s; a given vector $\mathbf{P}$ is directed along the face diagonal AC. Compute the moment of $\mathbf{P}$ about point O.

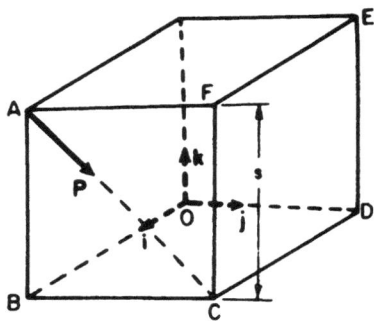

Fig. 18

We choose a set of rectangular coordinates with origin at O, this being the point about which the moment of $\mathbf{P}$ is sought. The position vector is $OA = OB + BA$, or

$$\mathbf{r} = s\mathbf{i} + s\mathbf{k} = s\,(\mathbf{i} + \mathbf{k}).$$

Recalling Theorem (6) of Section 8, i.e., $\mathbf{P} = (\mathbf{P} \cdot \mathbf{i})\mathbf{i} + (\mathbf{P} \cdot \mathbf{j})\mathbf{j} + (\mathbf{P} \cdot \mathbf{k})\mathbf{k}$, we find

$$\mathbf{P} = (P \cos 45)\mathbf{j} - (P \cos 45)\mathbf{k} = \frac{P}{\sqrt{2}}\,(\mathbf{j} - \mathbf{k}).$$

Hence,
$$\mathbf{M} = \mathbf{r} \times \mathbf{P} = \frac{sP}{\sqrt{2}}(-\mathbf{i} + \mathbf{j} + \mathbf{k}).$$

The magnitude of this vector is $M = sP\sqrt{\frac{3}{2}}$, the direction cosines of its line of action are $l = -\frac{1}{\sqrt{3}}$, $m = n = +\frac{1}{\sqrt{3}}$. (Note that this is the direction of the body diagonal BE.)

The unit vector parallel to **M** is
$$\mathbf{M_o} = \frac{\mathbf{M}}{M} = l\mathbf{i} + m\mathbf{j} + n\mathbf{k} = \frac{1}{\sqrt{3}}(-\mathbf{i} + \mathbf{j} + \mathbf{k}).$$

To find the moment of **P** about any line, L, passing through O, we perform the operation $\mathbf{M} \cdot \mathbf{L_o}$, where $\mathbf{L_o}$ is a unit vector in the direction of L. Thus, the moment of **P** about OD is $\mathbf{M} \cdot \mathbf{j} = sP/\sqrt{2}$; the moment of **P** about body diagonal OF is
$$\mathbf{M} \cdot (\mathbf{i} + \mathbf{j} + \mathbf{k})/\sqrt{3} = sP/\sqrt{6}.$$

Example 2: The line of action of a vector $\mathbf{Q} = 2\mathbf{j} + \mathbf{k}$ passes through point B whose coordinates with respect to origin O are $(-1, 1, 1)$. Find the moment of **Q** about line L which passes through O and through point C $(1, 2, 2)$ in the direction from O to C.

In order to compute the moment of **Q** about the line L, it is necessary first to find the moment of **Q** about point O on L. The appropriate position vector, from O to B (on **Q**), is (p. 10)
$$\mathbf{r} = (x_2 - x_1)\mathbf{i} + (y_2 - y_1)\mathbf{j} + (z_2 - z_1)\mathbf{k}$$
$$= -\mathbf{i} + \mathbf{j} + \mathbf{k},$$
so that
$$\mathbf{M} = \mathbf{r} \times \mathbf{Q} = -\mathbf{i} + \mathbf{j} - 2\mathbf{k}$$

The unit vector along L in the sense from O to C is
$$\mathbf{L_o} = l\mathbf{i} + m\mathbf{j} + n\mathbf{k}$$
$$= \tfrac{1}{3}(\mathbf{i} + 2\mathbf{j} + 2\mathbf{k});$$
hence, the moment of **Q** about L is
$$M_L = \mathbf{M} \cdot \mathbf{L_o} = -1.$$

Theorem of Varignon: The sum of moments about a point O (or about a line L passing through O) of vectors **P**, **Q**, **R**, ..., with common origin A, is equal to the moment about O (or about line L) of the single vector $(\mathbf{P} + \mathbf{Q} + \mathbf{R} + ...)$ representing their sum, with origin at A.

The proof of this theorem follows immediately from the distributive law of vector multiplication—Theorem (5) of Section 9.

At the beginning of this section we defined the moment M of a vector P about point O as the product $M = r \times P$ where r is the position vector extended from O to a point A on the line of action of P. We did not, however, specify which point A on P enters into the computation. We will now show that the location of A on P is immaterial, i.e., we will prove the following

Theorem: The moment of a vector about a point is unaltered by sliding the vector along its line of action.

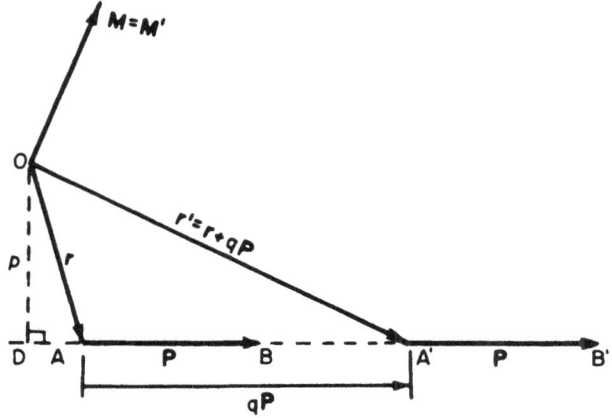

Fig. 19

In Fig. 19, AB represents the original position of P, so that the moment of P about O is $M = r \times P$. Now, let P be displaced along its line of action to position $A'B'$, so that the new moment of P about O is $M' = r' \times P$. But $OA' = OA + AA'$, and AA' must be expressible as a product of some scalar q and the vector P (see Section 5); hence, $r' = r + qP$. We have, then,

$$M' = r' \times P = (r + qP) \times P$$
$$= r \times P + qP \times P$$
$$= M \quad \text{since, by Theorem (2) of Section 9, } P \times P = 0.$$

This result might have been anticipated intuitively, since the magnitude of M is $rP \sin(r, P) = pP$, where $p = r \sin(r, P)$ represents the perpendicular distance OD from O to the line of action of P and this perpendicular distance does not change with the position of P along AB.

Corrolary: For two vectors which are of equal magnitude and have the same line of action but which are directed oppositely, the vector sum of moments about any point in space is zero.

The moment of a vector **P** about a line L was defined as the scalar component on L of the (vector) moment of **P** about a point O located on L. Suppose, however, we were to choose some other point, O', on L in the computation of the moment of **P** about L. How would the result be affected? As illustrated in Fig. 20, the moment of **P** about L may be expressed in terms of quantities referred to point O:

$$M_L = \mathbf{M}_O \cdot \mathbf{L}_o,$$

where $\mathbf{M}_O = \mathbf{r} \times \mathbf{P}$ and $\mathbf{L}_o$ represents a unit vector along line L measured in the positive direction.

If, now, the moment of **P** about L is expressed in terms of quantities referred to O', it becomes:

$$M'_L = \mathbf{M}_{O'} \cdot \mathbf{L}_o,$$

where $\mathbf{M}_{O'} = \mathbf{r'} \times \mathbf{P} \neq \mathbf{M}_O$.

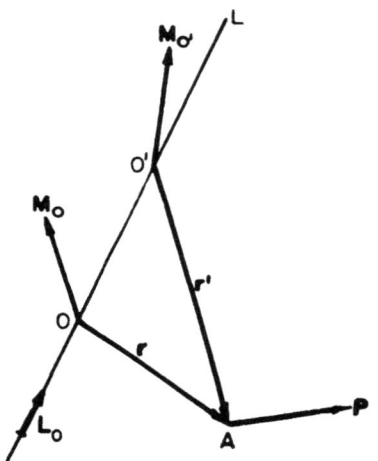

Fig. 20

We may relate the position vectors **r** and **r'** by noting that $OA = OO' + O'A$. But OO' is a scalar multiple of the unit vector $\mathbf{L}_o$, so that

$$\mathbf{r'} = \mathbf{r} - k\mathbf{L}_o$$

Hence,

$$\mathbf{M}_{O'} = (\mathbf{r} - k\mathbf{L}_o) \times \mathbf{P}$$

and

$$M'_L = [(\mathbf{r} - k\mathbf{L}_o) \times \mathbf{P}] \cdot \mathbf{L}_o$$
$$= (\mathbf{r} \times \mathbf{P}) \cdot \mathbf{L}_o - k(\mathbf{L}_o \times \mathbf{P}) \cdot \mathbf{L}_o$$

Consider the second term on the right. The cross product $\mathbf{L}_o \times \mathbf{P}$ represents a vector perpendicular to $\mathbf{L}_o$ (as well as to **P**); hence, its dot product with $\mathbf{L}_o$ must vanish identically since the two vectors are mutually perpendicular. Thus, we find that

$$M'_L = (\mathbf{r} \times \mathbf{P}) \cdot \mathbf{L}_o = M_L,$$

a result expressible as the

Theorem: The moment of a vector about a line is unaffected by the choice of the reference point on that line.

11. Differentiation of Vectors

Consider a vector **P** which is related to a scalar parameter s in such a manner that, as s varies continuously, so also does **P**. Then **P** is called a function of s, written $\mathbf{P} = \mathbf{P}(s)$.

Let, now, the parameter s be given an increment Δs as a result of which the vector function **P** acquires a corresponding increment $\Delta \mathbf{P}$. We then define the vector

$$\frac{d\mathbf{P}}{ds} = \lim_{\Delta s \to 0} \left[\frac{\mathbf{P}(s + \Delta s) - \mathbf{P}(s)}{\Delta s} \right]$$

$$= \lim_{\Delta s \to 0} \left(\frac{\Delta \mathbf{P}}{\Delta s} \right),$$

since $\mathbf{P}(s + \Delta s) = \mathbf{P} + \Delta \mathbf{P}$. The vector $\frac{d\mathbf{P}}{ds}$ is called the *derivative* of the vector **P** with respect to the scalar s.

We note that, in rectangular component form,

$$\frac{d\mathbf{P}}{ds} = \frac{dP_x}{ds}\mathbf{i} + \frac{dP_y}{ds}\mathbf{j} + \frac{dP_z}{ds}\mathbf{k},$$

i.e., the components of the derivative of a vector are equal to the derivatives of the respective components.

In analogous fashion we may define derivatives of higher order.

The following theorems derive from the definition of a derivative of a vector function:

(1) $\dfrac{d(c\mathbf{P})}{ds} = c\dfrac{d\mathbf{P}}{ds}$, where c is a constant.

(2) $\dfrac{d\mathbf{P}}{ds} = \dfrac{d\mathbf{P}}{dq}\dfrac{dq}{ds}$ (implicit differentiation).

(3) $\dfrac{d(\mathbf{P} + \mathbf{Q})}{ds} = \dfrac{d\mathbf{P}}{ds} + \dfrac{d\mathbf{Q}}{ds}$. The operation is linear.

(4) $\dfrac{d}{ds}[f(s)\mathbf{P}(s)] = f\dfrac{d\mathbf{P}}{ds} + \dfrac{df}{ds}\mathbf{P}$.

(5) $\dfrac{d(\mathbf{P} \cdot \mathbf{Q})}{ds} = \lim_{\Delta s \to 0} \dfrac{(\mathbf{P} + \Delta \mathbf{P}) \cdot (\mathbf{Q} + \Delta \mathbf{Q}) - \mathbf{P} \cdot \mathbf{Q}}{\Delta s}$

$= \dfrac{d\mathbf{P}}{ds} \cdot \mathbf{Q} + \mathbf{P} \cdot \dfrac{d\mathbf{Q}}{ds}$, since $\lim_{\Delta s \to 0} \left(\dfrac{\Delta \mathbf{P} \cdot \Delta \mathbf{Q}}{\Delta s} \right) = 0$.

ELEMENTS OF VECTOR ALGEBRA

(6) $\dfrac{d(\mathbf{P} \times \mathbf{Q})}{ds} = \dfrac{d\mathbf{P}}{ds} \times \mathbf{Q} + \mathbf{P} \times \dfrac{d\mathbf{Q}}{ds}$, where the order of the factors must be preserved.

(7) Let $\mathbf{P}_o(s)$ be a unit vector (this vector is a function of s by virtue of the fact that its direction changes, although, of course, its magnitude remains constant at unity); then $\mathbf{P}_o \cdot \mathbf{P}_o = 1$. Now, by Theorem (5) of this section,

$$\dfrac{d(\mathbf{P}_o \cdot \mathbf{P}_o)}{ds} = 2\mathbf{P}_o \cdot \dfrac{d\mathbf{P}_o}{ds} = \dfrac{d(1)}{ds} = 0.$$

Hence, $\mathbf{P}_o \perp \dfrac{d\mathbf{P}_o}{ds}$, by Theorem (4) of Section 8; that is, the derivative of a unit vector is always perpendicular to the unit vector. (It should be noted that the derivative of a unit vector is not, in general, a unit vector.)

As an illustration of the concept of derivative of a vector, we will consider the notion of velocity and acceleration expressed in a Cartesian system of coordinates. Let us suppose a point A to be moving in space along some curve C, and denote the position vector of A with respect to a *fixed* point O by $\mathbf{r}$ (Fig. 21). Thus, $\mathbf{r}$ is a function of the time t, and may be expressed in rectangular coordinates x, y, z, of point A with respect to O by

$$\mathbf{r}(t) = x(t)\mathbf{i} + y(t)\mathbf{j} + z(t)\mathbf{k}.$$

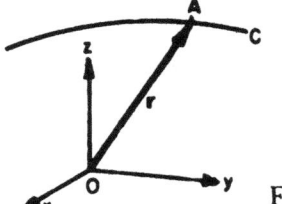

Fig. 21

We define the *velocity* of A relative to O as the vector

$$\mathbf{v} = \dfrac{d\mathbf{r}}{dt} = \dfrac{dx}{dt}\mathbf{i} + \dfrac{dy}{dt}\mathbf{j} + \dfrac{dz}{dt}\mathbf{k}.$$

In Newton's abbreviated notation, the derivative of a function with respect to time is denoted by a dot placed above the symbol denoting the function. Thus,

$$\mathbf{v} = \dot{\mathbf{r}} = \dot{x}\mathbf{i} + \dot{y}\mathbf{j} + \dot{z}\mathbf{k}.$$

This quantity may now be treated as any other vector in computing its magnitude and direction by means of expressions derived in Section 7.

ELEMENTS OF VECTOR ALGEBRA

Thus, e.g., the magnitude of the velocity vector is given by

$$v = \sqrt{\dot{x}^2 + \dot{y}^2 + \dot{z}^2}$$

and is commonly called the *speed*.

The *acceleration* of A with respect to O is defined as the vector

$$\mathbf{a} = \frac{d\mathbf{v}}{dt} = \frac{d^2\mathbf{r}}{dt^2};$$

expressed in rectangular component form, and in Newtonian notation,

$$\mathbf{a} = \dot{v}_x\,\mathbf{i} + \dot{v}_y\,\mathbf{j} + \dot{v}_z\,\mathbf{k} = \ddot{x}\,\mathbf{i} + \ddot{y}\,\mathbf{j} + \ddot{z}\,\mathbf{k}.$$

Example: The coordinates of a point are given as functions of the time by the expressions

$$x = \tfrac{1}{2} at^3 - bt^2 + ct + d,$$
$$y = \tfrac{1}{2} at^3 + ct,$$
$$z = \tfrac{1}{2} bt^2 + 2ct,$$

where a, b, c, d are constants. Compute the least magnitude of the acceleration and the time at which it occurs.

The velocity is

$$\begin{aligned}\mathbf{v} &= \dot{x}\,\mathbf{i} + \dot{y}\,\mathbf{j} + \dot{z}\,\mathbf{k} \\ &= (\tfrac{3}{2} at^2 - 2bt + c)\,\mathbf{i} + (\tfrac{3}{2} at^2 + c)\,\mathbf{j} + (bt + c)\,\mathbf{k}.\end{aligned}$$

The acceleration is

$$\begin{aligned}\mathbf{a} &= \ddot{x}\,\mathbf{i} + \ddot{y}\,\mathbf{j} + \ddot{z}\,\mathbf{k} \\ &= (3at - 2b)\,\mathbf{i} + 3at\,\mathbf{j} + b\,\mathbf{k}.\end{aligned}$$

The magnitude of the acceleration is

$$\begin{aligned}a = \sqrt{\ddot{x}^2 + \ddot{y}^2 + \ddot{z}^2} &= \sqrt{(3at - 2b)^2 + (3at)^2 + (b)^2} \\ &= \sqrt{18 a^2 t^2 - 12 abt + 5 b^2}.\end{aligned}$$

Hence, for least magnitude,

$$\frac{da}{dt} = \frac{36 a^2 t - 12 ab}{2\sqrt{18 a^2 t^2 - 12 abt + 5 b^2}} = 0,$$

or

$$t = \frac{b}{3a};$$

this is the time at which the least magnitude occurs. The value of the least magnitude is, then,

$$a_{\min} = \sqrt{18 a^2 \left(\frac{b}{3a}\right)^2 - 12 ab \left(\frac{b}{3a}\right) + 5 b^2}$$
$$= b\sqrt{3}.$$

Chapter II. THE PROBLEM OF EQUILIBRIUM

1. Introduction

It is desirable to start our discussion with several basic definitions:

Motion is the change of position with time.

A *particle* is a material body whose position may be represented adequately by a mathematical point and whose linear dimensions are sufficiently small so that they have a negligible effect on the description of motion (i.e., they are small compared with the other linear dimensions entering the particular problem).

A *material body* is an array of particles of physically small volume elements. In particular, a *rigid body* is an array of particles such that the distance between any pair of particles remains unchanged.

A *force* acting on a particle is conceived as an effort, of the nature of a push or pull, having a certain direction and a certain magnitude, i.e., having, mathematically, the nature of a vector.

A system is in *equilibrium* when its velocity is constant (in particular, zero).

We are now in a position to state what we mean by:

Mechanics, the study of motion of material bodies; this may be subdivided into:

  a) *Statics*, the study of the equilibrium of systems acted upon by forces.

  b) *Kinematics*, the study of possible motions of systems, without reference to the forces causing the motions.

  c) *Dynamics*, the study of actual motions of systems due to forces acting on them.

2. Composition of Forces and Moments of Forces

In accordance with our definition of force, we may apply all rules and theorems of vector algebra to the composition of forces. The validity of the definition is, of course, subject to verification by experiment. Further, since in most problems of statics we are concerned mainly with the relation of various forces with one another, the question of precise definition of units of measurement need not arise. A discussion of these is relegated to a course dealing with dynamics.

Two forces, **P** and **Q**, acting on a given particle, have a resultant **P** + **Q** which may be obtained graphically as shown in Fig. 22. The assertion that forces represented by AB, BC, CA are in equilibrium (i.e., their sum is zero) is called the *Triangle Law*.

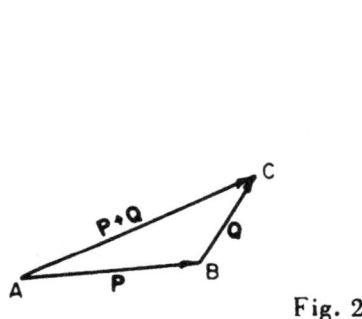
Fig. 22

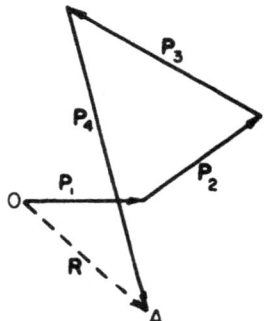
Fig. 23

To find the sum of more than two forces graphically, draw a polygon, as illustrated in Fig. 23. Thus, if the successive vectors are drawn so that their initial points coincide with the final points of the preceding vectors, their sum, or *resultant* (**R** in Fig. 23), is represented by the vector drawn from the initial point of the first to the final point of the last vector. This construction is given the name *force polygon*.

To find the resultant of several forces analytically, we recall that the components of the sum of a number of vectors are given by the sum of the respective components of the vectors. That is, if

$$\mathbf{R} = \mathbf{P}_1 + \mathbf{P}_2 + \ldots + \mathbf{P}_n \equiv \sum_{i=1}^{n} \mathbf{P}_i ,$$

then the components of **R** are given by

$$R_x = \sum_{i=1}^{n} P_{ix}, \quad R_y = \sum_{i=1}^{n} P_{iy}, \quad R_z = \sum_{i=1}^{n} P_{iz} .$$

Further, the magnitude and direction cosines of **R** are given by

$$R = \sqrt{R_x^2 + R_y^2 + R_z^2}, \quad l = R_x/R, \quad m = R_y/R, \quad n = R_z/R.$$

In two-dimensional (plane) systems, $\gamma = 90°$; hence, $n = 0$ and $R_z = 0$. Therefore,

$$R = \sqrt{R_x^2 + R_y^2}, \quad \tan(\mathbf{R}, \mathbf{i}) \equiv \tan\theta = R_y/R_x .$$

THE PROBLEM OF EQUILIBRIUM

We may extend the correspondence between a force (physical concept) and a vector (mathematical concept) to moment of a force and moment of a vector. Thus, the moment of a force **F** about a given point O is expressed by the vector product $\mathbf{M} = \mathbf{r} \times \mathbf{F}$, where **r** is the position vector of **F** with respect to O. Physically, the moment of a force (or, for the sake of brevity, moment) represents an effort tending to turn a body about a line (the line of action of **M**), just as a force represents an effort tending to displace a body along a straight line.

The moment may be expressed in terms of its rectangular components. Thus, if $\mathbf{r} = x\mathbf{i} + y\mathbf{j} + z\mathbf{k}$ and $\mathbf{F} = F_x\mathbf{i} + F_y\mathbf{j} + F_z\mathbf{k}$, then

$$\mathbf{M} = M_x\mathbf{i} + M_y\mathbf{j} + M_z\mathbf{k},$$

where $M_x = (yF_z - zF_y)$, $M_y = (zF_x - xF_z)$, $M_z = (xF_y - yF_x)$.
In two-dimensional systems, $z = 0$ and $F_z = 0$; hence, $M_x = M_y = 0$ and the moment reduces to

$$\mathbf{M} = M_z\mathbf{k} = (xF_y - yF_x)\mathbf{k}.$$

We see that the total moment in this case is directed parallel to the z-axis and its magnitude is equal to the moment about the z-axis.

Still considering two-dimensional systems, we refer to Fig. 24. The moment of **F** about O has been shown to be $x_A F_y - y_A F_x$ in magnitude and directed parallel to the positive z axis, i.e., pointing out of the plane of the paper. We note that the expression for the magnitude consists of the sum of the products of each component and its (perpendicular) distance to a coordinate axis, provided that counterclockwise turning efforts are reckoned positive, clockwise turning efforts negative. In this process the observer is assumed placed along the positive z-axis looking toward the origin. We note, further, that

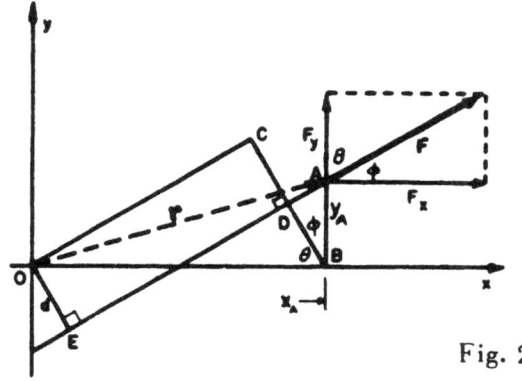

Fig. 24

$$M_z = x_A F_y - y_A F_x = F\left(x_A \frac{F_y}{F} - y_A \frac{F_x}{F}\right)$$
$$= F(x_A \cos\theta - y_A \cos\phi).$$

24 THE PROBLEM OF EQUILIBRIUM

But $x_A \cos \theta = CB$ and $y_A \cos \phi = DB$, so that $x_A \cos \theta - y_A \cos \phi = CD = OE$; that is,

$$M_z = Fd,$$

where F is the magnitude of the force and d is the (perpendicular) distance (called the *moment arm*) from the origin to the line of action of the force. The algebraic sign of the moment will depend on whether the turning effort is counterclockwise (positive) or clockwise (negative).

This result may be derived directly from the definition of a moment as a vector product. If, in Fig. 24, the position vector from O to A is denoted by $\mathbf{r}$, the magnitude of the moment of $\mathbf{F}$ about O is given by

$$|\mathbf{M}| = |\mathbf{r} \times \mathbf{F}|$$
$$= rF \sin \angle OAE = Fd.$$

Although Fig. 24 represents a two-dimensional system, this demonstration and result are equally valid for systems in space.

Example 1: Compute the magnitude and direction of the resultant of the forces depicted in Fig. 25.

It is convenient to arrange the computation in tabular form.

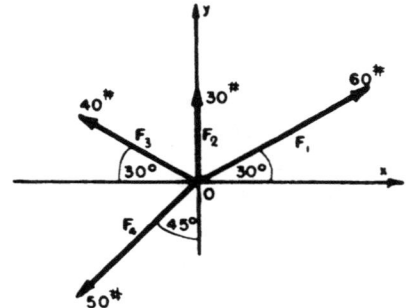

Fig. 25

Force	F_x	F_y
F_1	$60 \cos 30°$ = 51.96	$60 \cos 60°$ = 30
F_2	0	30
F_3	$-40 \cos 30°$ = -34.64	$40 \cos 60°$ = 20
F_4	$-50 \cos 45°$ = -35.36	$-50 \cos 45°$ = -35.36
	$\Sigma F_x = -18.04$	$\Sigma F_y = +14.64$

THE PROBLEM OF EQUILIBRIUM

$$R = \sqrt{(\Sigma F_x)^2 + (\Sigma F_y)^2} = 48.15 \text{ lbs.}$$
$$\theta_x = (\mathbf{R}, \mathbf{i}) = \arctan (\Sigma F_y)/(\Sigma F_x) = \arctan (-2.474)$$
$$= 112°.$$

The moment of each force about the origin O is, clearly, equal to zero.

Example 2: (a) Compute the magnitude and direction of the resultant of the forces shown in Fig. 26. (b) Compute the total moment of these forces about point A.

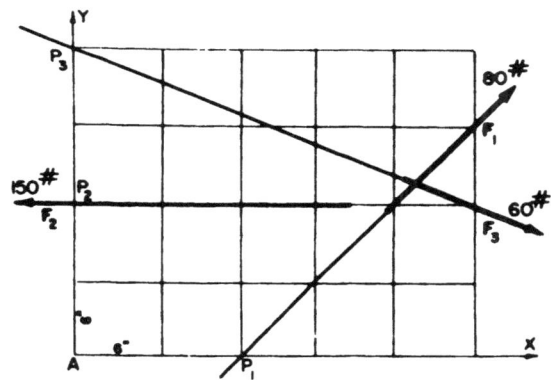

Fig. 26

Force	F_x	F_y
F_1	$80 \cos 45°$ = 56.57	$80 \cos 45°$ = 56.57
F_2	-150	0
F_3	$60(5/\sqrt{29})$ = 55.71	$-60(2/\sqrt{29})$ = -22.28
	$R_x = -37.72$	$R_y = +34.29$

Therefore, $R = \sqrt{(-37.72)^2 + (34.29)^2} = 50.98$ lbs.

$$\tan \theta_x = R_y/R_x = -0.9091; \quad \theta_x = 137°44'.$$

To compute the moments of individual forces about point A, we choose, in each case, the most convenient position vector. Thus, for F_1, we resolve this force into components at point P_1, since at P_1 the horizontal (x) component of F_1 has zero moment about A. Accordingly,

THE PROBLEM OF EQUILIBRIUM

$$M_1 = + (56.57 \text{ lb.})(12") = + 56.57 \text{ ft. lbs.}$$
Similarly, $\quad M_2 = + (150)(12") = + 150 \text{ ft. lbs.}$
and $\quad M_3 = - (55.71)(24") = -111.42 \text{ ft. lbs.}$

Hence, the total moment about A is, in magnitude,

$$M_A = M_1 + M_2 + M_3 = + 95.15 \text{ ft. lbs.} ;$$

the plus sign indicates that the resultant moment has a counterclockwise turning effort, i.e., the moment vector is directed along the positive z-axis.

Example 3: (a) Find the magnitude and direction of the resultant of the forces shown in Fig. 27.

$$\mathbf{F}_1 = -40\mathbf{i} \text{ (lbs.)}$$
$$\mathbf{F}_2 = -20\mathbf{j}$$
$$\mathbf{F}_3 = 60(8/10)\mathbf{i} + 60(6/10)\mathbf{k}$$
$$\quad = 48\mathbf{i} + 36\mathbf{k}$$

$$\mathbf{R} = 8\mathbf{i} - 20\mathbf{j} + 36\mathbf{k}$$
$$R = \sqrt{(8)^2 + (-20)^2 + (36)^2} = 42 \text{ lbs.}$$
$$l = 8/42, \quad \alpha = 79°$$
$$m = -20/42, \quad \beta = 118°25'$$
$$n = 36/42, \quad \gamma = 31°.$$

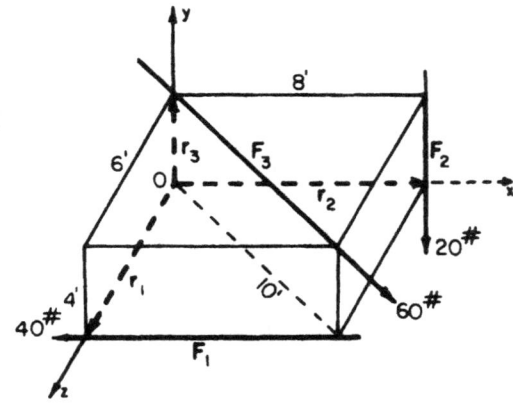

Fig. 27

(b) Find the magnitude and direction of the resultant moment about point O of the forces shown in Fig. 27.

The most convenient position vectors are drawn in Fig. 27. Thus,

$\mathbf{r}_1 = 6\mathbf{k}$ (ft.); $\quad \mathbf{M}_1 = \mathbf{r}_1 \times \mathbf{F}_1 = 6\mathbf{k} \times (-40\mathbf{i}) = -240\mathbf{j}$ (ft. lbs.)
$\mathbf{r}_2 = 8\mathbf{i} \quad\quad\quad ; \quad \mathbf{M}_2 = \mathbf{r}_2 \times \mathbf{F}_2 = -160\mathbf{k}$
$\mathbf{r}_3 = 4\mathbf{j} \quad\quad\quad ; \quad \mathbf{M}_3 = \mathbf{r}_3 \times \mathbf{F}_3 = -192\mathbf{k} + 144\mathbf{i}$

Hence,
$$\mathbf{M} = \sum_{i=1}^{3} \mathbf{M}_i = 144\mathbf{i} - 240\mathbf{j} - 352\mathbf{k}$$

$M = \sqrt{(144)^2 + (-240)^2 + (-352)^2} = 450$ ft. lbs.
$l' = 144/450, \quad m' = -240/450, \quad n' = -352/450;$
$\alpha' = 71°20', \quad \beta' = 122°15', \quad \gamma' = 141°30'.$

3. Newton's Laws for a Particle

The foundation of the science of theoretical mechanics rests on three postulates, set forth by Newton (1687). Their validity, not demonstrable mathematically, is based exclusively on experimental evidence. That is, an essentially affirmative answer may be given to the question: Do the theoretical calculations, based on the three postulates, predict satisfactorily experimental results? The statements are:

(1) Every body continues in its state of rest or of uniform motion in a straight line, except insofar as it may be compelled by force to change that state.

(2) Change of motion is proportional to the applied force, and takes place in the direction of the straight line in which the force acts.

(3) To every action there is always an equal and contrary reaction; or, equivalently, the mutual actions of any two bodies are always equal and oppositely directed.

The second statement is the Law of Motion which may be stated mathematically as follows:

$$\mathbf{F} = \frac{d}{dt}(m\mathbf{v}) = m\mathbf{a},$$

where m (assumed here to be constant) is the *mass* of the particle, defined as a measure of resistance to motion. $\mathbf{F}$ is the applied force, $\mathbf{v}$ and $\mathbf{a}$ are the velocity and acceleration with respect to a fixed origin.

The third statement is called the Law of Action and Reaction. The first statement, called the Law of Inertia, is a special case of the Law of Motion.

4. Equilibrium

The statement of the Law of Inertia, given in the preceding section, refers to a state of rest (zero velocity) or of uniform motion in a straight line (constant velocity). This is precisely our definition of equilibrium (see Section 1 of this chapter). Accordingly, starting with Newton's second law, we may formulate the conditions for equilibrium of a particle.

Consider, first, our particle in a state of equilibrium; i.e., its velocity, v = constant. Therefore, its acceleration, a = 0, so that, according to the second law, F = 0. We see that, in order for equilibrium to exist, it is *necessary* that F = 0.

Now, consider the converse, i.e., suppose F = 0. Hence, according to the second law, a = 0 and, therefore, v is constant or zero; that is, the particle is in equilibrium. We see that, in order for equilibrium to exist, it is *sufficient* that F = 0.

We may state these results in the form of the following

Theorem: A necessary and sufficient condition for a particle to be in a state of equilibrium is that the vector sum (resultant) of all forces acting on it be equal to zero.

In component form F = 0 becomes

$$F_x = 0, \quad F_y = 0, \quad F_z = 0.$$

Graphically this means that the force polygon must close.

Our discussion to this point has referred to a single particle. On the other hand, in a system of particles, we distinguish between:

Internal forces: forces on a particle of a system exerted by another particle of that system. These forces always occur in pairs, being equal and oppositely directed (in accordance with Newton's third law).

External forces: forces on a particle of a system exerted by an agency outside that system.

It should be noted in this connection that internal forces may be converted into external ones simply by excluding a certain portion from the system under consideration. Conversely, by adding to a system, external forces may be converted into internal ones. This will be discussed more fully in the next chapter.

Let us now consider, for the sake of simplicity, a system consisting of three particles, A, B, and C (Fig. 28). Let **P**, **Q**, **R** represent the resultant external force acting on A, B, C, respectively, and let **U**, **V**, **W** represent internal forces acting, respectively, between the pairs of particles AB, BC, AC; for example, **U** denotes the force exerted on A by B, etc. Now, for the system to be in equilibrium, each particle must be in equilibrium. But, for equilibrium of each particle, the sum of all

THE PROBLEM OF EQUILIBRIUM

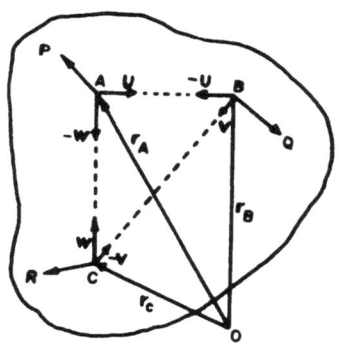

Fig. 28

forces acting on it must equal zero. Referring to Fig. 28, we find

$$F_A = P + U - W = 0$$
$$F_B = Q - U + V = 0$$
$$F_C = R - V + W = 0$$
$$\Sigma F = F_A + F_B + F_C = P + Q + R = 0$$

The procedure may be stated in words:

(a) The (vector) sum of all forces acting on all particles is zero.
(b) The (vector) sum of all internal forces is zero, since these occur in pairs each of which consists of equal and oppositely directed forces.
(c) Hence, the (vector) sum of all *external* forces acting on the system is zero. This is a necessary condition for equilibrium of a system.

In the preceding discussion we did not take account of the fact that each pair of internal forces is such that the forces are not only equal and oppositely directed, but collinear as well (as drawn in Fig. 28). This is, of course, a consequence of Newton's Law of Action and Reaction. Taking cognizance of this fact leads to an additional necessary condition. We have already seen that (referring to Fig. 28)

$$F_A = 0, \quad F_B = 0, \quad F_C = 0;$$

therefore, for an arbitrary point O, $r_A \times F_A = 0$, $r_B \times F_B = 0$, $r_C \times F_C = 0$, and, consequently, $r_A \times F_A + r_B \times F_B + r_C \times F_C = 0$.

This may be written

$$r_A \times P + r_B \times Q + r_C \times R +$$
$$(r_A - r_B) \times U + (r_B - r_C) \times V + (r_C - r_A) \times W = 0.$$

But $r_A - r_B$ and U are collinear vectors, so that their vector product is zero. The same holds for each of the other terms in the second line of this equation. Hence,

$$r_A \times P + r_B \times Q + r_C \times R = 0.$$

Again, the argument may be stated in words: In order for a particle to be in equilibrium, the resultant of all forces acting on it must be zero; hence, the moment of this resultant about an arbitrary point is zero. A similar argument holds for all particles. Thus,

THE PROBLEM OF EQUILIBRIUM

(a′) The (vector) sum of moments about an arbitrary point of all forces acting on the system is zero.

(b′) But the (vector) sum of moments of all internal forces is zero, since the (vector) sum of moments of a pair of equal, collinear and oppositely directed forces is zero.

(c′) Hence, the (vector) sum of moments of all *external* forces is zero.

Conditions (c) and (c′) constitute *necessary* conditions for the equilibrium of a system of particles. They may be written, symbolically, in vector form

$$\mathbf{F} = 0, \quad \mathbf{M} = 0,$$

or, in terms of components, in scalar form

$$F_x = 0, \quad F_y = 0, \quad F_z = 0$$
$$M_x = 0, \quad M_y = 0, \quad M_z = 0.$$

These conditions are also *sufficient* to insure equilibrium but the proof of this assertion will not be demonstrated (see Chapter VII, Section 4 for additional discussion).

The truth of the following theorem may be easily ascertained:

Theorem: The action of a given system of forces will in no way be changed if we add to or subtract from these forces another system of forces in equilibrium. This statement is called the *principle of superposition*.

The principle of superposition leads directly to another theorem of great importance. Consider a rigid body C under the action of force **P** acting as shown in Fig. 29(a). Now add the system **P′**, **P″**, shown in Fig. 29(b); this is such that **P′** = −**P″** = **P**, so that **P′** and **P″** form a system of forces in equilibrium. By the principle of superposition the effect

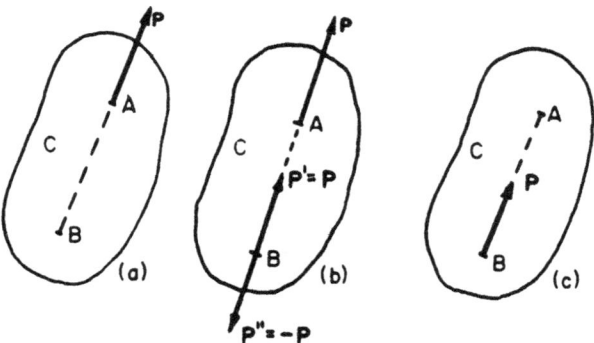

Fig. 29

on the body of the system of forces shown in Fig. 29(b) is the same as that shown in Fig. 29(a). Now remove the system (in equilibrium) **P, P″**. Again, this operation is without effect on the original system. We are left, therefore, with force **P′** (= **P**) applied at B instead of at A, as shown in Fig. 29(c). Thus, we have proved the following

Theorem: The point of application of a force may be transmitted along its line of action without changing the effect of the force on any *rigid* body to which it may be applied. This is called the *principle of transmissibility of force*.

Chapter III. EQUILIBRIUM OF SIMPLE PLANAR SYSTEMS

1. Introduction

Plane statics deals with systems in equilibrium under the action of coplanar forces. The plane of action (taken to be the xy plane) is called the *fundamental plane*. Accordingly, the number of conditions of equilibrium, when written in component form, reduces to three:

$$F_x = 0, \quad F_y = 0, \quad M_z = 0.$$

We recall that the first two equations require the vanishing of two mutually perpendicular components of the resultant of all *external* forces acting on the system, while the third equation demands that the resultant moment of *external* forces about any axis normal to the fundamental plane be zero.

2. Free-Body Diagram

As discussed in Chapter II and reiterated in the preceding section, the conditions of equilibrium deal exclusively with external forces. In most problems of statics, however, we wish to evaluate internal forces acting within a system. Hence, we seek a way of transforming internal into external forces by isolating, judiciously, portions of the original system. To this end we employ the *free-body diagram*—this is simply the sketch of a body removed from contact with all other bodies but with all external forces shown acting on it.

Let us compile a catalogue of the types of reactions which occur frequently, together with the corresponding free-body diagrams. It should be noted that the weight (W) of a body is an external force, taken to be acting through the center of gravity of the body (for uniform bodies the center of gravity coincides with the geometric center). The direction of the vector representing the weight is always vertically downward.

a) *Smooth Floor* (Fig. 30): reaction is compressive and always directed normal to the common tangent line at the point of contact.

EQUILIBRIUM OF SIMPLE PLANAR SYSTEMS

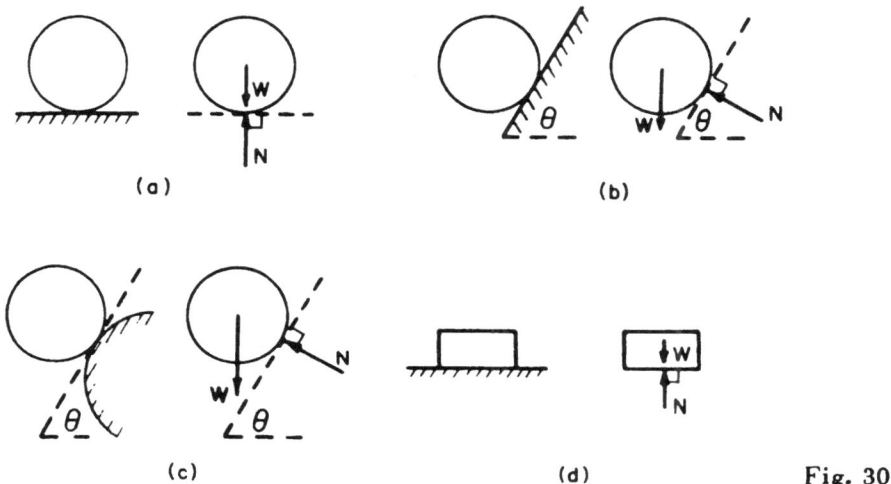

Fig. 30

b) *Smooth Knife Edge* (Fig. 31): reaction is compressive and directed normal to the body at the point of contact.

Fig. 31

c) *Roller* (Fig. 32): reaction is compressive and directed normal to the common tangent line at the point of contact.

Fig. 32

d) *Roller Nest* (Fig. 33): reaction is compressive and directed normal to the base of the rollers. However, the use of this symbol is commonly extended to include the possibility of a tensile reaction.

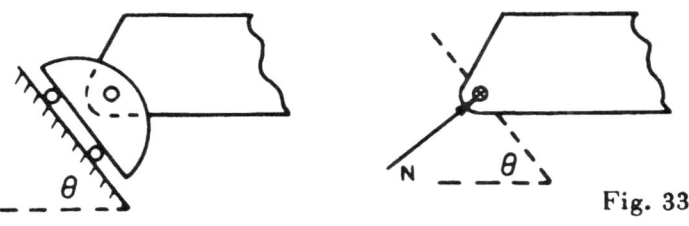

Fig. 33

34 **EQUILIBRIUM OF SIMPLE PLANAR SYSTEMS**

e) *Flexible Cord* (Fig. 34): reaction is tensile and collinear with the cord.

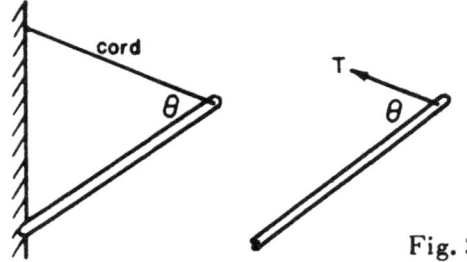

Fig. 34

f) *Smooth Pin* (Fig. 35): both the magnitude and the direction of the reaction are unknown. It is convenient, in this case, to replace the resultant reaction by its coordinate components, the direction of the latter being chosen arbitrarily.

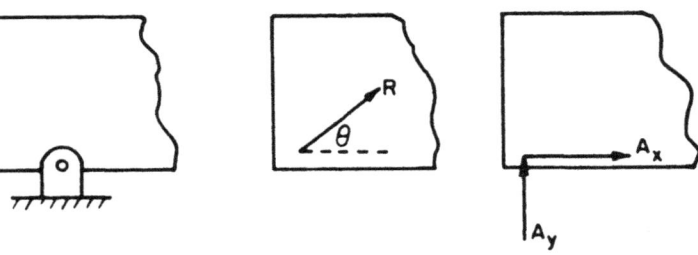

Fig. 35

3. Concurrent Forces

The simplest type of problem concerns systems each of which can be analyzed with the aid of free-body diagrams which contain only concurrent forces, i.e., forces whose lines of action meet at a single point. This is best illustrated by examples.

Example 1: A ball O, suspended from string AC, rests against the smooth vertical wall AB, as shown in Fig. 36(a). If the angle BAC between the string and the wall is α and the weight of the ball is W, compute the magnitude of the tension, T, in the string and of the contact force, N, between the ball and the wall.

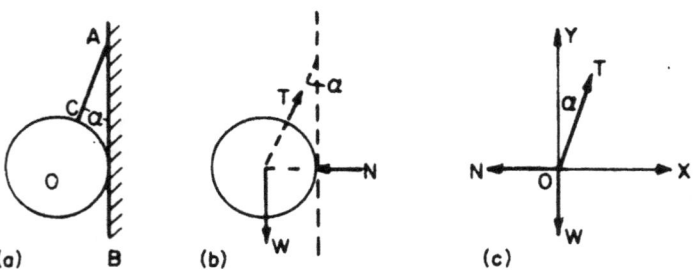

Fig. 36

EQUILIBRIUM OF SIMPLE PLANAR SYSTEMS

The tension, T, and the contact force, N, are internal forces in the system comprising the ball, the string and the wall. A free-body diagram of the ball (Fig. 36b) serves to convert T and N into external forces. Finally, Fig. 36c shows the diagram of forces all of which are considered to be acting at their common point O. (This is an application of the principle of transmissibity of force discussed in Section 4 of Chapter II.)

Setting the sum of forces in the coordinate directions equal to zero, we find the system will be in equilibrium if, simultaneously,

$$\Sigma F_x = T \sin \alpha - N = 0,$$
$$\Sigma F_y = T \cos \alpha - W = 0.$$

These are two algebraic equations on the two unknown quantities T and N. Their solution is

$$T = W/\cos \alpha, \quad N = W \tan \alpha.$$

We note that the equation of moments about the z-axis is satisfied identically. This is always true in concurrent force problems and is a result of Varignon's theorem (Chapter I, Section 10). We note, further, that the solution is independent of the size of the ball.

Example 2: A ball O, weighing 12 lbs., lies in contact with two mutually perpendicular smooth planes AB and BC. Find the force against each surface if plane BC is inclined 60° to the horizontal, as shown in Fig. 37.

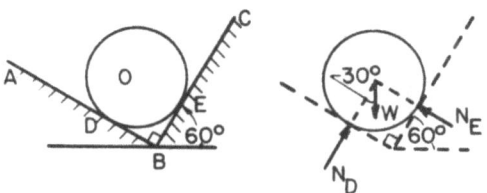

Fig. 37

The conditions of equilibrium lead to the equations

$$\Sigma F_x = N_D \cos 60 - N_E \cos 30 = 0,$$
$$\Sigma F_y = N_D \sin 60 + N_E \sin 30 - W = 0,$$

the solution of which is $N_E = W/2 = 6$ lbs. and $N_D = W\sqrt{3}/2 = 10.4$ lbs. The forces exerted on the planes at D and E have, of course, directions opposite to those shown acting on the ball O.

We could have simplified our computational task by choosing more suitable coordinate axes. Thus, since N_D and N_E are perpendicular to each

36   **EQUILIBRIUM OF SIMPLE PLANAR SYSTEMS**

other, we may take an x' axis along N_D and a y' axis along N_E. The conditions of equilibrium now read

$$\Sigma F_{x'} = N_D - W \cos 30 = 0,$$
$$\Sigma F_{y'} = N_E - W \cos 60 = 0;$$

these lead immediately to the solution given above.

Example 3: Two smooth cylinders, A (weighing 40 lbs. and having a diameter of 16 in.) and B (weighing 30 lbs. and having a diameter of 10 in.), are placed in a box as shown in Fig. 38. Find the reactions at C, D, E and F.

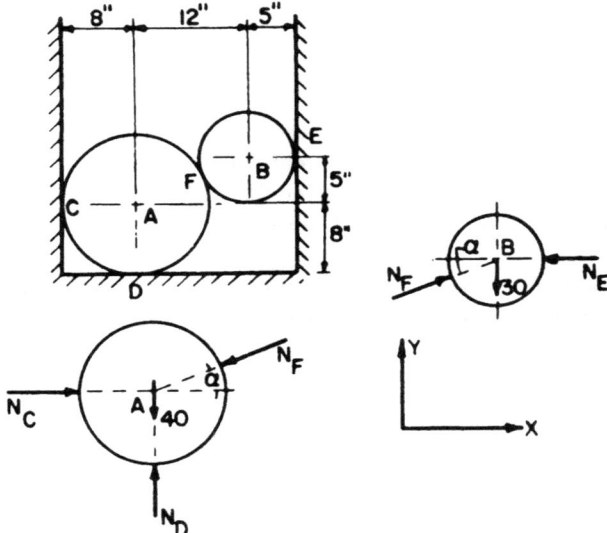

Fig. 38

A free-body diagram of A contains three unknown forces, N_C, N_D, N_F, for the solution of which only two equations of equilibrium are available. Accordingly, we seek additional independent equations which we obtain from consideration of the equilibrium of B. The free-body diagram of B introduces one more unknown force, N_E, but supplies two more equations of equilibrium. It should be noted that, as a result of the Law of Action and Reaction, the forces at F exerted on A and B are equal in magnitude and oppositely directed. The solution now proceeds as in previous examples:

body A: $\Sigma F_x = N_C - N_F \cos \alpha = 0,$
                     $\Sigma F_y = N_D - N_F \sin \alpha - 40 = 0;$

body B: $\Sigma F_x = N_F \cos \alpha - N_E = 0,$
                     $\Sigma F_y = N_F \sin \alpha - 30 = 0.$

EQUILIBRIUM OF SIMPLE PLANAR SYSTEMS

But, from Fig. 38 we see that $\cos \alpha = 12/13$ and $\sin \alpha = 5/13$, so that $N_F = 78$ lbs., $N_C = N_E = 72$ lbs., $N_D = 70$ lbs.

4. Parallel Forces

In this case the equation of force equilibrium in the direction perpendicular to the common direction of the forces is satisfied identically. Consequently, in problems involving parallel coplanar force systems, we need consider only one force and one moment equation.

Example: A beam 24 ft. long carries a load at each end and is supported as shown in Fig. 39. Find the reactions at the supports, (a) neglecting

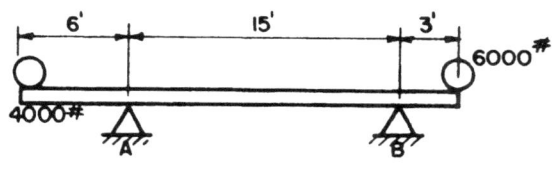

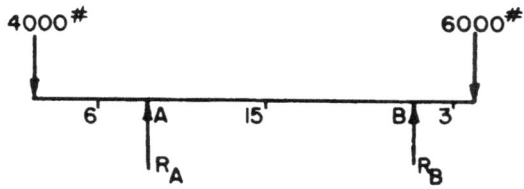

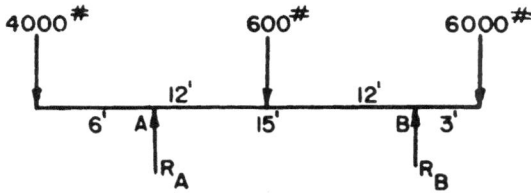

Fig. 39

the weight of the beam, (b) taking into account that the beam is uniform and weighs 25 lbs. per ft.

(a) The free-body diagram is shown in the second of the diagrams in Fig. 39. The force equation of equilibrium is

$$\Sigma F_y = R_A + R_B - 4000 - 6000 = 0.$$

Since the sum of moments must be zero about *any* point in the plane, we choose the most convenient one, i.e., one which yields an equation of moment equilibrium containing the least possible number of unknown quantities. Such a point is A. Thus, reckoning counterclockwise moments positive and clockwise moments negative, we find

$$\Sigma M_A = 4000(6) + R_B(15) - 6000(18) = 0.$$

Hence, R_B = 5600 lbs., R_A = 4400 lbs.
(b) Since the beam is uniform, its weight of (25 lb/ft) (24 ft.) = 600 lbs. is taken to act at the geometric center, as shown in the bottom sketch of Fig. 39. The equations now become

$$\Sigma F_y = R_A + R_B - 4000 - 600 - 6000 = 0,$$
$$\Sigma M_A = 4000(6) - 600(6) + R_B(15) - 6000(18) = 0,$$

and their solution is R_B = 5840 lbs., R_A = 4760 lbs.

The solution may equally well have been arrived at by employing a second moment equation in place of the force equation. It must be emphasized, however, that for any given planar free-body diagram, the number of *independent* scalar equations does not exceed three and these include equations which are satisfied identically. Further discussion is deferred until Section 6 of this chapter and Section 3 of Chapter IV.

5. General Coplanar Forces

In this case the three equations of equilibrium (Section 1 of this chapter) must be applied explicitly.

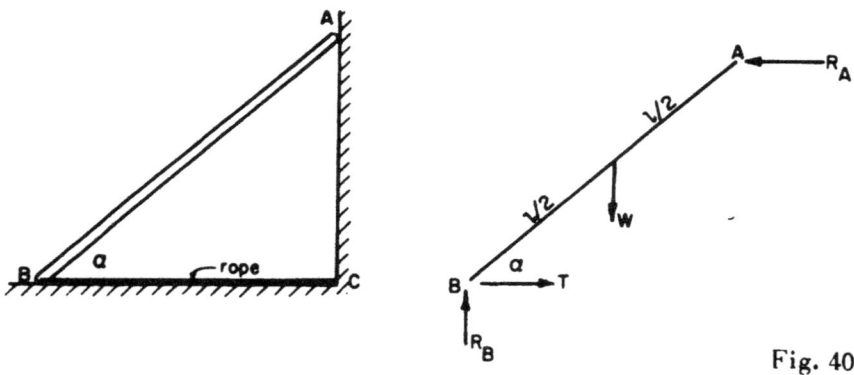

Fig. 40

Example: A uniform ladder AB, of weight W and length l, rests against a smooth vertical wall and on a smooth horizontal floor. It is restrained from sliding by a horizontal rope BC fastened to the wall, such that angle ABC is α, as shown in Fig. 40. Find the reactions on the wall and floor, and the tension in the rope.

EQUILIBRIUM OF SIMPLE PLANAR SYSTEMS

Choosing coordinate axes in the usual directions, we may write the equations of equilibrium:

$$\Sigma F_x = T - R_A = 0,$$
$$\Sigma F_y = R_B - W = 0,$$
$$\Sigma M_B = R_A (l \sin \alpha) - W \left(\frac{l}{2} \cos \alpha\right) = 0.$$

The solution of these is

$$T = R_A = W/2 \tan \alpha, \quad R_B = W;$$

we note that the reactions are independent of the length of the ladder.

6. Static Determinateness and Indeterminateness

We shall say of a problem that it is statically *determinate* if the number of unknown quantities is equal to the number of independent equations of equilibrium (excluding those equations which are satisfied identically). However, many problems arise in which the principles of statics are inadequate to furnish a unique solution. Such problems are referred to as *indeterminate*. As an example, consider a horizontal beam resting on three supports and subjected to the action of its own weight W (Fig. 41).

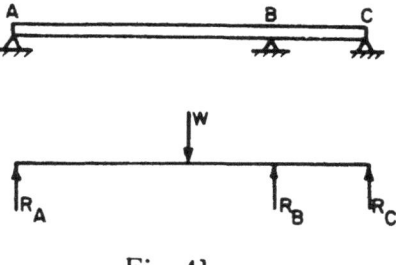

Fig. 41

The equilibrium of the horizontal force components is satisfied identically, so that we are left with two independent equations for the determination of three unknown reactions. The invocation of additional moment equations will not yield additional information; this will be demonstrated in Section 3 of the following chapter. It would, however, be incorrect to state that a solution does not exist, since, in an actual case, the reaction at each support will have a definite value.

The difficulty is due to the assumption, made implicitly, that the beam is rigid; if this assumption is removed, that is, if, in the analysis, the beam is permitted to deform, the difficulty is found to be removed as well. To elaborate on this point, suppose we have constructed a model in which one of the supports is movable in a direction normal to the axis of the beam, say by means of a screw placed at C. At the start, let the screw just clear the beam, so that $R_C = 0$. We may now compute the values of R_A and R_B by applying the principles of statics alone; the corresponding set of values of the reactions will be R'_A, R'_B, 0. Subsequent raising of the level of C will cause the beam to bend, and the values of the reactions will change: R_A will increase, R_B decrease, R_C increase (from zero). Suppose the level of C to be raised until the contact at B is broken. Again, the principles of statics will yield a solution, and the values of the reactions will now be R''_A, 0, R''_C. We have, then, a set of values of the reactions which varies continuously between R'_A, R'_B, 0 and R''_A, 0, R''_C, each of the intermediate, as well as the terminal, values corresponding to positions of equilibrium of the system. That is, the values of the three reactions depend on the relative heights of the supports and the elastic properties of the beam.

Chapter IV. EQUIVALENCE OF FORCE SYSTEMS

1. Introduction

Two systems of forces are called *statically equivalent* or *equipollent* when the following conditions are satisfied:

(a) the (vector) sum of all the forces of one system is equal to the (vector) sum of all the forces of the other system;

(b) the (vector) sum of the moments of all the forces of one system about an arbitrary point in space is equal to the (vector) sum of the moments of all the forces of the other system about the same point.

Stated in symbols, the conditions for static equivalence of two force systems are

$$\mathbf{F'} = \mathbf{F''}, \quad \mathbf{M'} = \mathbf{M''}.$$

If, for any system of forces, $\mathbf{F} = 0$, $\mathbf{M} = 0$ (i.e., the system is in equilibrium), the system is said to be equipollent to zero.

In the case of two-dimensional force systems, the coordinate system being chosen such that the xy plane is the plane of the forces, the conditions of equivalence, when expressed in terms of the scalar components, become:

$$F'_x = F''_x, \quad F'_y = F''_y, \quad M'_z = M''_z.$$

Example: Find the magnitude, direction, and location of the resultant of the system of parallel forces shown in Fig. 42.

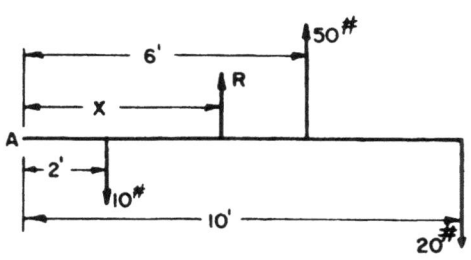

Fig. 42

Clearly, the resultant, **R**, of the forces is parallel to the forces. Its direction is indicated by the algebraic sign of the sum. Thus, setting the resultant equal to the sum of the forces and taking the upward direction as positive, we find

$$R = 50 - 20 - 10 = + 20 \text{ lbs. } up.$$

42   EQUIVALENCE OF FORCE SYSTEMS

To find the location of the line of action of **R**, we must set the moment of **R** about a line normal to the plane of the figure and passing through *any* point in the plane equal to the resultant moment of all the forces about the same line. Thus, considering moments about an axis through point A, we find

$$xR = -2(10) + 6(50) - 10(20) = +80 \text{ ft. lbs.},$$

or
$$x = \frac{80}{R} = 4 \text{ ft. from } A \text{ (to the right)}.$$

2. Couples

A system of two parallel forces of equal magnitude but acting in opposite directions constitutes a *couple*. Referring to Fig. 43, we see that a couple produces a moment about any point in space, O, equal to

$$\mathbf{M} = \mathbf{r} \times (-\mathbf{P}) + \mathbf{r}' \times (\mathbf{P})$$
$$= -\mathbf{r} \times \mathbf{P} + (\mathbf{r} + \mathbf{s}) \times \mathbf{P}$$
$$= \mathbf{s} \times \mathbf{P}.$$

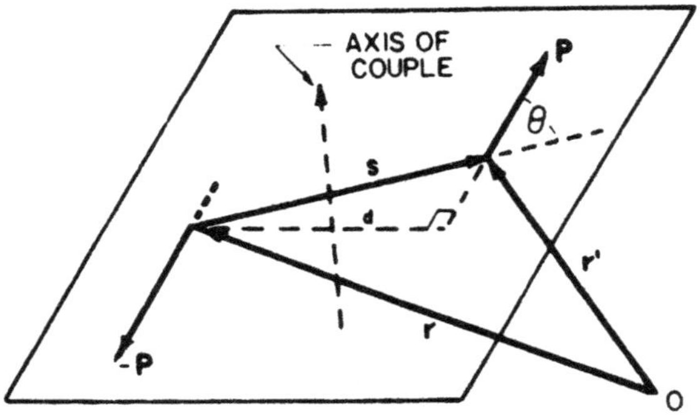

Fig. 43

Thus, the magnitude of the moment, $sP \sin \theta$, is equal to the product of the magnitude of one of the forces, P, and the (perpendicular) distance, d (called the *arm* of the couple), between the lines of action of the two forces. We see, further, that the magnitude of the moment of a couple is independent of the point O, that is, the moment of a couple is the same for all points in space. The couple exerts a turning effort about a line (called the *axis* of the couple) perpendicular to the plane of the two forces constituting the couple. The positive direction of the axis is determined by the right-hand rule. It is seen that a couple is characterized by (1) the magnitude of its moment and (2) the direction of its axis; i.e., its properties are those of a vector. It should be noted, further, that a couple cannot be resolved into a single force—it is an irreducible quantity.

Finally, since two couples which have the same (vector) moment comprise statically equivalent systems ($F' = F'' = 0$, $M' = M''$), a couple is completely specified by its moment. In fact, we may think of a couple as a pure moment. Hence, when we speak of a couple G, we have in mind any one of an infinite number of couples each of which has a moment G.

In coplanar force systems, the axis of a couple is perpendicular to the fundamental plane. The moment of a couple is reckoned positive if the couple tends to produce counterclockwise rotation, negative if the couple tends to produce clockwise rotation, as illustrated in Fig. 44.

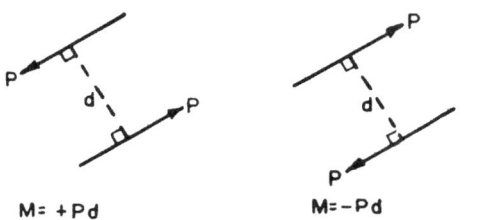

M = +Pd   M = −Pd   Fig. 44

3. Reduction of a System of Forces

A system of forces whose resultant is **F** and whose resultant moment about some point A is **M** is statically equivalent to the system consisting of a single force **F**, applied at A, and a couple **M**. This follows from the definition of static equivalence, and leads directly to the

Theorem: Any system of forces is statically equivalent to a single force, applied at an arbitrary point, together with a couple.

Example 1: Reduce the system of forces shown in Fig. 27 to a single force, applied at O, and a couple.

The conditions for static equivalence will be met if the single force at O is equal to the resultant of all forces of the system in Fig. 27 and if the moment of the couple is equal to the resultant moment of all the forces in Fig. 27 about O. Accordingly, making use of the results of Example 3, pp. 26-27, we find

$$\text{the single force at } O: \mathbf{R} = 4(2\mathbf{i} - 5\mathbf{j} + 9\mathbf{k}),$$
$$\text{the couple:} \qquad \mathbf{G} = 16(9\mathbf{i} - 15\mathbf{j} - 22\mathbf{k}).$$

The requirement to effect a reduction of the original system to a single force, applied at some other point, A, and a couple is met by the same force **R** as given above, but the couple is now equal to the resultant moment of the original force system about A. If the resultant moment about

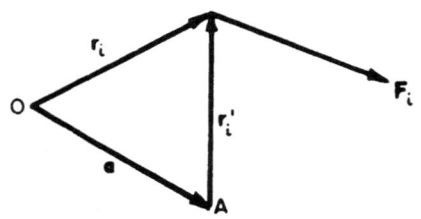

Fig. 45

O is already known, then the moment about A may be obtained immediately in terms of M_O. Referring to Fig. 45, let F_i represent a typical force of the system of n forces, and let r_i and r_i' be its position vectors with respect to O and A, respectively. Hence, a being the position vector of A with respect to O,

$$M_A = \sum_{i=1}^{n} r_i' \times F_i = \sum_{i=1}^{n} (r_i - a) \times F_i$$
$$= \sum_{i=1}^{n} r_i \times F_i - a \times \sum_{i=1}^{n} F_i$$
$$= M_O - a \times R,$$

where R is the resultant of all the forces of the system.

It may be seen from this result that, if the resultants of two systems of forces are equal and if the resultant moments about a given point O are equal, then the resultant moments about any other point A will be equal. In particular, if for a given system $R = 0$, $M_O = 0$ (the system is equipollent to zero, i.e., in equilibrium), then the resultant moment M_A about any point A will vanish identically. We have here proof of the assertions, made in Sections 4 and 6 of the preceding chapter, that, for a given planar system, there are no more than three independent (nontrivial) scalar equations of equilibrium. In three dimensions, the number of such equations does not exceed six.

A system of forces may also be reduced to an equivalent system consisting of two forces, each of whose lines of action passes through a specified point. Another reduction is possible which results in a single force, localized along a unique line of action, and a couple whose axis is parallel to the line of action of this force. This is called a *canonical* system and is distinguished by the property that the couple appearing in it represents the smallest possible couple. The proof is omitted here.

For a system of forces in a single plane, a further reduction is possible. Suppose the original system to be reduced to a single force F at O and a couple N (chosen, for the sake of illustration, counterclockwise), as shown in Fig. 46. The couple may be replaced by the system $-F$ at O and F at A, such that $Fd = N$. Since the forces at O are now in equi-

EQUIVALENCE OF FORCE SYSTEMS

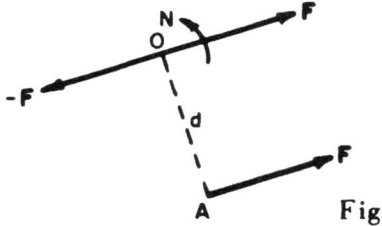

Fig. 46

librium, the original system is statically equivalent to a single force **F** applied at A. It should be noted that, in this case, A is not a previously specified point; it must be determined by calculation. Further, if $\mathbf{F} = 0$, the system reduces to a couple. These results may be stated as a

Theorem: Any force system in the plane may be reduced either to a single force or, if the force resultant is zero, to a couple.

Example 2: Reduce the system of forces shown in Fig. 26 to a single force and find the points of intersection of its line of action with the coordinate axes.

It was previously established, in Example 2 on pp. 25-26, that the resultant force is of magnitude $R = 50.98$ lbs. and its inclination to the positive x-axis is $\theta_x = 137°44'$. The resultant moment about A is counterclockwise and has a magnitude $M_A = 95.15$ ft. lb. Hence, the line of action of the single force will be a distance $l = M_A/R = 1.866$ ft. from A, as shown in Fig. 47. It is seen that $AP_1 = l/\cos(\theta_x - 90) = 2.774$ ft. and $AP_2 = l/\sin(\theta_x - 90) = 2.522$ ft.

It should be noted that a major reason for a discussion of static equivalence is the use of this concept in the reduction of a general force system to a simple one, thus facilitating the analysis of the effect of the force system on a rigid body.

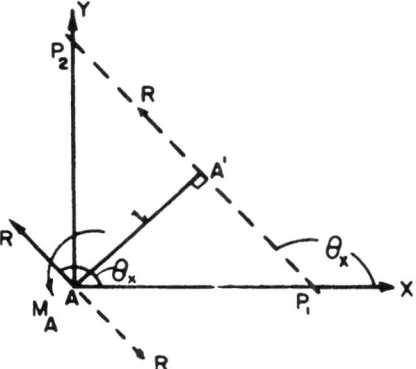

Fig. 47

Chapter V. SIMPLE STRUCTURES

1. Trusses

A *truss* is a structural entity consisting of a network of bars fastened at their ends (joints). This type of structure, illustrated in Fig. 48, is widely used in bridges, in roof supports, and in similar instances where the loading can be applied at discrete points of the structure. Exterior members are called *chords*, interior ones are called *web* members. The

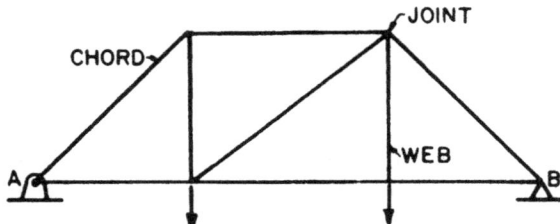

Fig. 48

members may be made of steel and fastened by rivets, or of wood and held together by bolts. We shall confine our attention to trusses with joints that are coplanar and subjected to coplanar loading. In order to simplify the analysis, it is customary to make the following assumptions:

(a) The bars are rigid, and their weight is negligible in comparison to the applied forces.

(b) The joints act like smooth hinges, so that each bar is capable of rotation about either end without any resisting moment.

(c) External loads imposed on the truss are applied exclusively at the joints.

Assumptions (a) and (c) imply that all forces, external and reactive, acting on a given bar are applied at its joints. In order that the condition of force equilibrium be satisfied, the resultants of all forces at each joint must be equal in magnitude and oppositely directed, as shown in Fig. 49. Now, by virtue of assumption (b), the joints are incapable of exerting a resisting moment; hence, in order for the condition of moment equilib-

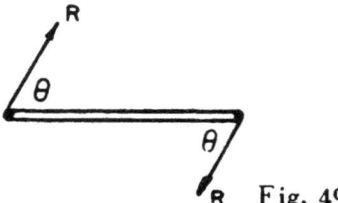

Fig. 49

rium to be fulfilled, the moment of each reaction in Fig. 49 about the opposite joint must vanish. Since **R** is, in general, different from zero, this is possible only if $\theta = 0$. Thus, the reaction is collinear with the bar, and each bar will be subjected to either pure tension or pure compression along its axis, but not to bending. Tension occurs when the forces exerted by the joints on a bar act away from each other, compression when they act toward each other (Fig. 50a). Of course, if a diagram is drawn of the reactions exerted by a bar on a joint, then, in accordance with the Law of Action and Reaction, the forces will be directed oppositely from those shown in Fig. 50a (see Fig. 50b).

Finally, our analysis will be confined to trusses which are statically determinate, i.e., such that they can be analyzed by applying conditions of statics alone. The problem consists of finding the magnitude and sense of the forces in each bar of a truss which result from the application of given external loading.

In Fig. 48 joint A is fixed, while joint B is constrained to move in a horizontal direction. We note that the truss is fixed by these conditions and behaves as a rigid body. Moreover, it is *just-rigid*, since removal of one of its members destroys its rigidity. Suppose a truss has b bars connected at j joints. If the coordinates of the joints on a bar of length l are (x_1, y_1) and (x_2, y_2), then the rigidity of the bar is expressed by the equation

$$(x_2 - x_1)^2 + (y_2 - y_1)^2 = l^2.$$

Since there are b bars, there are b relations of this type connecting the $2j$ coordinates. Further, since we have imposed, as a result of the external fixity, three more conditions (e.g., zero displacement in the horizontal and vertical directions at A, zero displacement in the vertical direction at B, Fig. 48), for a just-rigid truss $b + 3$ conditions determine the $2j$ coordinates of the joints, i.e., fix the whole structure. Hence, for a just-rigid truss,

$$b = 2j - 3.$$

The problem of analyzing a just-rigid truss involves finding the magnitude of the forces in b bars and of the three support reactions, a total of

$b + 3$ unknown quantities. Consequently, $2j$ independent equations of equilibrium are required in order to attain static determinateness. As will be shown in Section 2 below, two independent equations are available at each joint. Hence, the relation $b + 3 = 2j$ insures that the truss is statically determinate as well as just-rigid.

The smallest number of joints which will create a just-rigid truss is three. In this case, $b = 3$, so that we have a triangular truss (Fig. 51a).

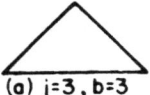 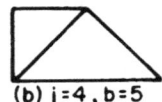 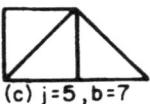 Fig. 51

Examples of just-rigid trusses in which $j > 3$ are given in Figs. 51 b, c. One way to build up a just-rigid truss is to start with a triangular form and add two concurrent bars at a time. Trusses constructed in this manner are termed *simple*, but the method does not furnish all just-rigid trusses (see Section 4, below).

If, in a structure, $b < 2j - 3$, the structure is said to be *under-rigid*: it will collapse when subjected to load. On the other hand, if a truss is constructed such that $b > 2j - 3$, it is said to be *over-rigid* and is statically indeterminate. The additional members are said to be *redundant* (Fig. 52).

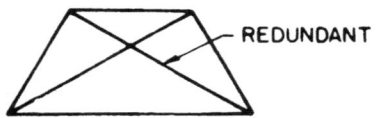

Fig. 52

2. The Method of Joints

In the analysis of a truss by the method of joints, each joint is considered as a particle in equilibrium under the action of external loads (if any) and of the (internal) reactions of members at the joint. Mathematically, then, we are faced with a problem of concurrent forces. Thus, after the external reactions (of supports) have been computed by considering the entire structure as a free body, we proceed by starting at any joint at which there are no more than two unknown bar reactions,

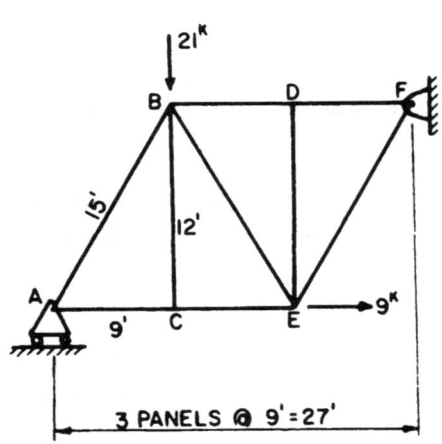

Fig. 53

SIMPLE STRUCTURES

solve for these, and then continue to an adjacent joint. In this manner the forces in the entire truss may be found.

Example: Find the forces in the members of the truss shown in Fig. 53. (The superscript k on the numbers representing external loading denotes the unit of load 1 kip = 1000 lbs.)

To compute the external reactions, we refer to a free-body diagram of the entire truss (Fig.

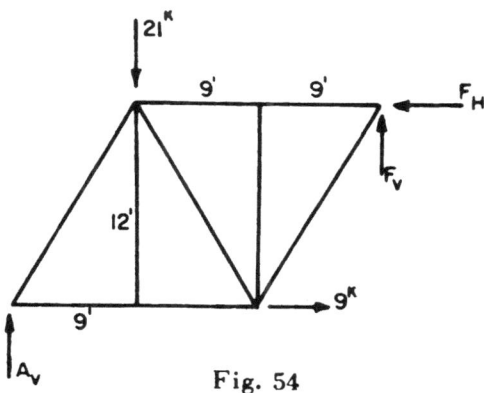

Fig. 54

54). Applying the conditions of equilibrium to this structure, we find

$$\Sigma F_x = 9 - F_H = 0;$$
$$\therefore F_H = 9.$$
$$\Sigma F_y = A_V - 21 + F_V = 0.$$
$$\Sigma M_F = -27 A_V + 21 \times 18 + 9 \times 12 = 0;$$
$$\therefore A_V = 18 \uparrow$$
$$F_V = 3 \uparrow.$$

We may note at this point that rollers are employed at joint A (Fig. 53) not only to make the problem statically determinate, but to provide opportunity for movement resulting from changes in length caused by changes in temperature or due to sagging of the support at F. Prevention of such movement would give rise to forces of large magnitudes.

Only two bars meet at A, so that this joint affords a convenient starting point for our analysis. Consider all forces acting on the joint at A, assuming the directions of AB and AC as shown in Fig. 55. Hence,

$$\Sigma F_y = A_V - AB \times \tfrac{12}{15} = 0;$$
$$\therefore AB = 22.5 C.$$

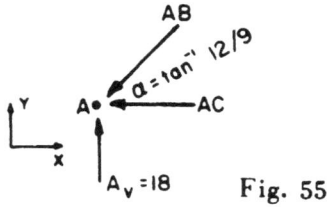

Fig. 55

The value of AB is positive, indicating that the direction of the force is correct as assumed. Since AB pushes on pin A, the pin pushes on the member; hence, AB is in compression, denoted by the letter C following its numerical value.

$$\Sigma F_x = -AC - AB \times \tfrac{9}{12} = 0;$$
$$AC = -13.5\,T.$$

The negative sign indicates that the direction shown in Fig. 55 is incorrect. AC pulls on the pin; hence, the member is in tension (denoted by the letter T).

Having evaluated the stresses in AC and AB, we move on to an adjacent joint at which no more than two unknown reactions appear. In the present case, such is joint C, but not B. Care must be taken to draw each *known* force in the correct direction in the free-body diagram of a subsequent joint. The analysis proceeds as follows:

Joint C

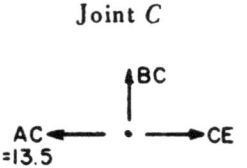

$\Sigma F_x = -13.5 + CE = 0$
$\therefore CE = 13.5\,T$
$\Sigma F_y = BC = 0$

Joint B

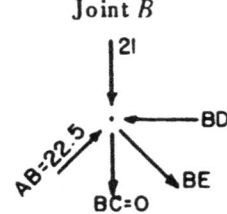

$\Sigma F_y = 22.5 \times \tfrac{12}{15} - BE \times \tfrac{12}{15} - 21 = 0$
$\therefore BE = -3.75\,C$
$\Sigma F_x = 22.5 \times \tfrac{9}{15} + (-3.75) \times \tfrac{9}{15} - BD = 0$
$\therefore BD = 11.25\,C$

Joint D

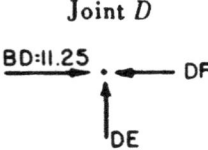

$\Sigma F_x = 11.25 - DF = 0$
$\therefore DF = 11.25\,C$
$\Sigma F_y = DE = 0$

Joint E

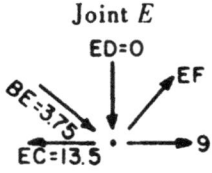

$\Sigma F_y = -3.75 \times \tfrac{12}{15} + EF \times \tfrac{12}{15} = 0$
$\therefore EF = 3.75\,T$

SIMPLE STRUCTURES

We may supplement our solution with a check. Thus, at F,

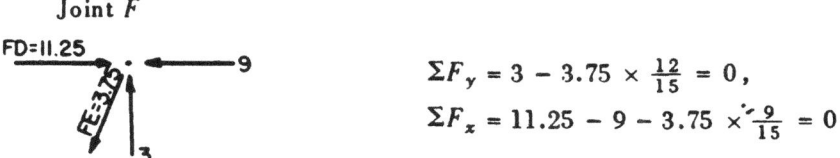

$$\Sigma F_y = 3 - 3.75 \times \tfrac{12}{15} = 0,$$
$$\Sigma F_x = 11.25 - 9 - 3.75 \times \tfrac{9}{15} = 0.$$

Some members with zero load are easily located before starting a joint by joint analysis. At an *unloaded* joint at which *three* members meet of which two are *collinear*, the third member will be unloaded no matter what angle it makes with the collinear members (Fig. 56). This is seen by applying the force equation of equilibrium, at the joint in question, in the direction perpendicular to the collinear members. In Fig. 56, such considerations with reference to joint C indicate that $BC = 0$.

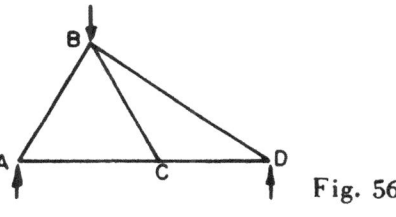

Fig. 56

If, on the other hand, a joint, such as the one discussed above, is *loaded*, the force in the non-collinear member may be found by inspection. Thus, referring to Fig. 57, $BC = P \sec \alpha$ (C). In particular, in a configuration such as shown in Fig. 58, $BD = W$ (C).

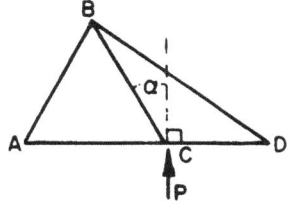

Fig. 57

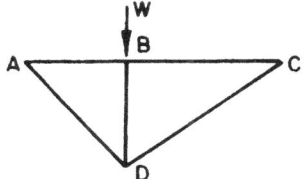

Fig. 58

3. The Method of Sections

The method of sections enables us to find the force in a given member without having to resort to a time-consuming joint-by-joint analysis. We imagine the member cut by a *section*, cutting as few other members as possible and in no case more than a total of three, but severing the truss into two portions. We then draw a free-body diagram of the portion of the truss to either side of the section, showing all external loads and previously determined supporting reactions, as well as the unknown forces at the cut members. What this procedure accomplishes is nothing more nor

52   SIMPLE STRUCTURES

less than the conversion of the unknown internal force in the bar into an external force by means of a properly chosen and isolated portion of the truss. The use of the equations of statics completes the procedure.

Example: Employ the method of sections to determine the force in member BE of the truss of Fig. 53.

A section is passed through BE and also through BD and CE, thus cutting the truss into two parts, but cutting no more than three members in the process. For the sake of illustration, we consider the free-body diagram of the portion of the truss to the left of the section (Fig. 59). We may now write three equations of equilibrium for the determination of the three unknown bar reactions. However, it is more convenient to set up one equation involving only one unknown. Thus, to find BE,

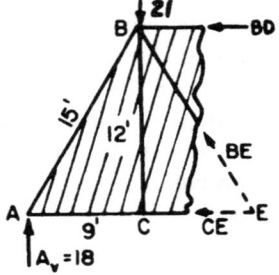

Fig. 59

$$\Sigma F_y = 18 - 21 + BE \times \tfrac{12}{15} = 0;$$
$$\therefore BE = 3.75\,C.$$

To find BD, for example, the appropriate equation is

$$\Sigma M_E = -18 \times 18 + 21 \times 9 + 12\,BD = 0;$$
$$\therefore BD = 11.25\,C.$$

To find CE, we use the equation

$$\Sigma M_B = -18 \times 9 - 12\,CE = 0,$$

from which
$$CE = -13.5\,T.$$

As a general rule it may be stated that the appropriate equation for the determination of a given force will be the one expressing the vanishing of the moments of all forces about the point of intersection of the other two bars severed by the section. In the event the latter are parallel, a force equation in the direction normal to the parallel bars will secure the answer.

4. Complex Trusses

Thus far we have limited our discussion to simple trusses, i.e., statically determinate frameworks which could be constructed by adding successive pairs of members to an initial triangular form. There are, however, trusses which cannot be constructed in this manner, although the condition for static determinateness, $b = 2j - 3$, is satisfied. We refer to these as *complex*. An example is shown in Fig. 60 in which members AD, BE and CF cross without touching and the truss is symmetric about the vertical line CF.

We note that the method of sections cannot be used in this case since a section cannot be passed cutting no more than three nonconcurrent members and severing the truss into two parts. The method of joints cannot be employed in the usual manner since there is no starting point available at which only two members meet.

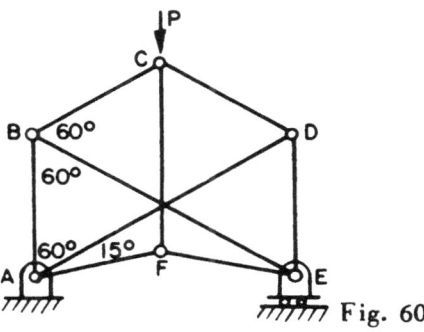
Fig. 60

We may proceed to analyze the truss of Fig. 60 by setting up two equations at each joint which express the conditions of force equilibrium. Since there are six joints in the truss, we will be confronted with twelve simultaneous, algebraic equations containing the nine unknown bar reactions and the three unknown support reactions. This, however, is much too tedious a procedure. Alternatively, we may employ the usual joint-by-joint procedure, expressing all bar reactions in terms of a single (unknown) one. On closing the calculation, this unknown reaction becomes determined. Thus, let us denote the force in CF by S, assumed tensile. Then,

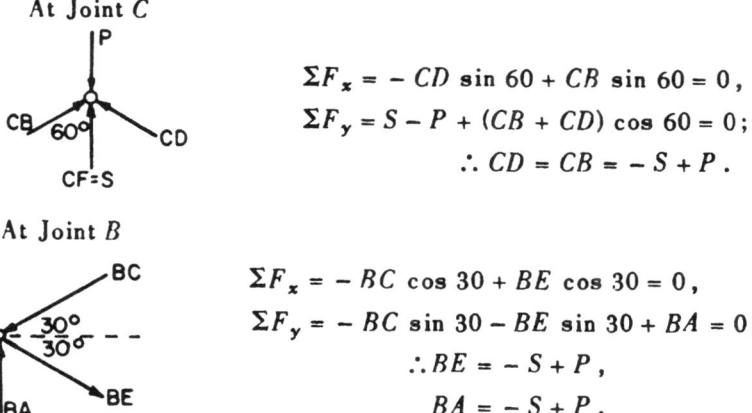

At Joint C

$$\Sigma F_x = - CD \sin 60 + CB \sin 60 = 0,$$
$$\Sigma F_y = S - P + (CB + CD) \cos 60 = 0;$$
$$\therefore CD = CB = - S + P.$$

At Joint B

$$\Sigma F_x = - BC \cos 30 + BE \cos 30 = 0,$$
$$\Sigma F_y = - BC \sin 30 - BE \sin 30 + BA = 0;$$
$$\therefore BE = - S + P,$$
$$BA = - S + P.$$

At Joint A

$$\Sigma F_x = -AF \cos 15 + AD \cos 30 = 0,$$
$$\Sigma F_y = R_A - AB + AD \sin 30 - AF \sin 15 = 0.$$

But $AD = BE$ by symmetry, so that

$$AF = AD \frac{\cos 30}{\cos 15} = (-S + P) \frac{\cos 30}{\cos 15}.$$

Moreover, the support reaction, obtained by considering equilibrium of the entire truss as a rigid body, is $R_A = P/2$. Hence,

$$\frac{S}{2} - (-S + P) \cos 30 \tan 15 = 0,$$

so that

$$S = \frac{2P \cos 30 \tan 15}{1 + 2 \cos 30 \tan 15} = 0.317 P.$$

The reactions may now be found; thus,

$$CB = CD = 0.683 P \ (C)$$
$$BE = DA = 0.683 P \ (T)$$
$$BA = DE = 0.683 P \ (C)$$
$$AF = FE = 0.612 P \ (C)$$
$$CF = 0.317 P \ (C)$$

As a check, consider Joint F:

$$\Sigma F_x = FA \cos 15 - FE \cos 15 = 0,$$
$$\Sigma F_y = FA \sin 15 + FE \sin 15 - S = 0;$$
$$\therefore FA = FE = S/2 \sin 15 = 0.612 P.$$

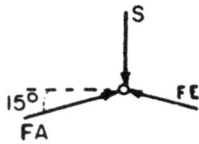

5. Frames

A pin-connected *frame* is distinguished from a truss by two characteristics: (a) the members need no longer be joined at their ends, and (b) the loads may be applied at any point of the structure instead of at the joints exclusively. As a result, the reactions in the members of a frame are, in general, no longer purely axial, but consist of forces which tend to produce bending of the members in addition to tension or compression. The members designed to withstand bending are of sturdier proportions than those in a truss and their weight is no longer neglected in the calculation of internal reactions.

SIMPLE STRUCTURES

Example 1: The crane shown in Fig. 61 is made of uniform beams, weighing 40 lb./ft., joined by smooth pins at C, E and F. The total weight of the crane and its load $W = 2000$ lbs., applied at D, is supported at A. Compute the reactions at A and B and the pin reactions at C, E and F.

It is convenient to consider the free-body diagram of the entire crane (Fig. 62). The weights of the individual members are shown acting at the geometric centers of the members. The equations of equilibrium are

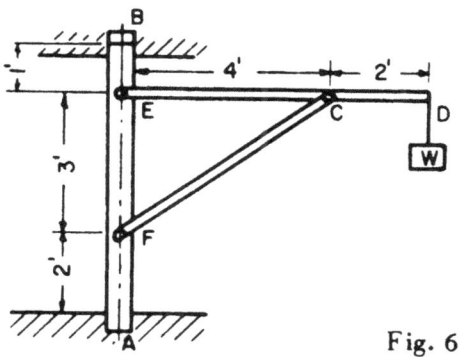

Fig. 61

$\Sigma F_x = A_H - B_H = 0$,
$\Sigma F_y = A_V - 240 - 200 - 240 - 2000 = 0$,
$\Sigma M_A = 6 B_H - 6 \times 2000 - 3 \times 240$
$\qquad\qquad - 2 \times 200 = 0$,

and their solution is

$A_H = B_H = 2187$ lbs.,
$A_V = 2680$ lbs.

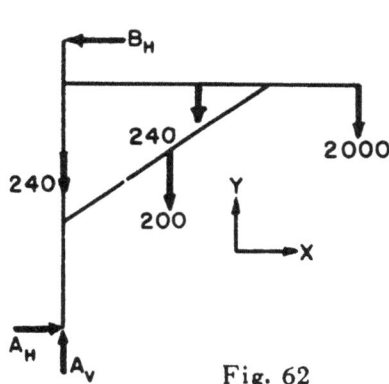

Fig. 62

Hence, the magnitude of the reaction at A is

$R_A = \sqrt{A_H^2 + A_V^2} = 3459$ lbs.

and its line of action is inclined to the x-axis by an angle

$\theta_x = \text{arc tan} \dfrac{A_V}{A_H} = \text{arc tan } 1.225 = 50°47'$.

There are two unknown components of the reaction at each of the three pins C, E and F. Accordingly, we must set up six independent equations of equilibrium involving these components. This is done by considering free-body diagrams of any two of the members, say of ED and FC. Consider ED first: assuming the reactions directed as shown in Fig. 63, we find the equations of equilibrium

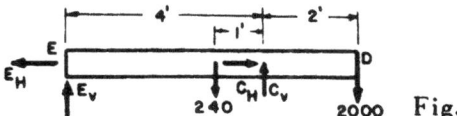

Fig. 63

56 SIMPLE STRUCTURES

$$\Sigma F_x = C_H - E_H = 0,$$
$$\Sigma F_y = E_V + C_V - 240 - 2000 = 0,$$
$$\Sigma M_E = -3 \times 240 + 4 C_V - 6 \times 2000 = 0,$$

from which

$$C_V = 3180 \text{ lbs. } \uparrow,$$
$$E_V = -940 \text{ lbs. } \downarrow,$$
$$C_H = E_H.$$

We now come to the free-body diagram of FC, shown in Fig. 64. The forces acting on FC at C are equal and opposite to the forces acting on ED at C. The equations of equilibrium may be written in the form

$$\Sigma F_x = F_H - C_H = 0,$$
$$\Sigma F_y = F_V - 200 - C_V = 0,$$
$$\Sigma M_F = 3 C_H - 4 C_V - 2 \times 200 = 0,$$

so that

$$C_H = E_H = F_H = 4373 \text{ lbs.},$$
$$F_V = 3380 \text{ lbs.}$$

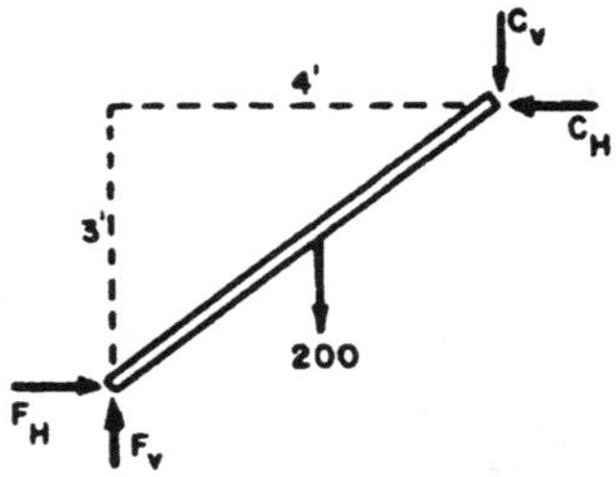

Fig. 64

Summary of pin reactions:

At E on bar ED: $E_H = 4373 \leftarrow$ $R_E = 4473$
                     $E_V = 940 \downarrow$ $\theta_x = 192°08'$

At C on bar ED: $C_H = 4373 \rightarrow$ $R_E = 5407$
                     $C_V = 3180 \uparrow$ $\theta_x = 36°02'$

At F on bar FC: $F_H = 4373 \rightarrow$ $R_F = 5527$
                     $F_V = 3380 \uparrow$ $\theta_x = 37°42'$

Example 2: Fig. 65a illustrates a portion of a frame which is subjected to a load W transmitted by means of a pulley C. It is desired to compute the reaction exerted by the pulley on bar AC at C. For the sake of simplicity we assume the pulley to be of negligible weight. The corresponding free-body diagram of the pulley, shown in Fig. 65b, leads to the equations

$$\Sigma M_c = -WR + TR = 0,$$
$$\Sigma F_x = -T \cos \alpha + C_H = 0,$$
$$\Sigma F_y = -W - T \sin \alpha + C_V = 0,$$

SIMPLE STRUCTURES

whose solution is

$T = W$ (as expected),
$C_H = W \cos \alpha$,
$C_V = W (1 + \sin \alpha)$.

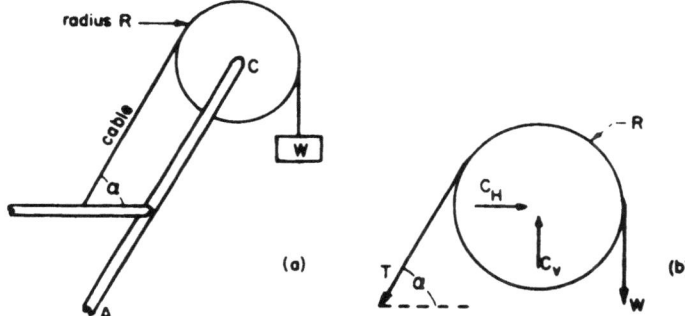

Fig. 65

Chapter VI. SLIDING FRICTION

1. Introduction

In Chapter III we noted that, in a *smooth* contact, a surface exerts a compressive reaction which is always directed along the normal to the surface at the contact. This property serves, in fact, to define such a contact. In actual cases, however, the reaction at a contact will, in general, be directed obliquely to the surface; the surfaces are said to be in *rough* contact. It is convenient to resolve the contact reaction into a (compressive) normal component, called the *normal force*, and a tangential component, termed the *force of friction*.

2. The Laws of Friction

We shall deal specifically with frictional forces which arise as a result of a tendency of two bodies in contact to slide with respect to one another. Hence, we speak of *sliding* (as opposed, for example, to *rolling*) friction. Further, we assume the contact to be devoid of a lubricating film (as in a bearing), so that the term *dry* friction is also applied to the phenomenon under consideration.

The actual physical processes which are responsible for the presence of sliding friction, that is, the mechanisms which create the tangential resistance to motion, are still imperfectly understood, and, although many theories have been proposed, no one has been accepted as definitive. It will suffice to state certain general results obtained experimentally.

The usual empirical laws of sliding friction are associated with the name of Coulomb (1785) although credit for some of them must go to Amontons (1699) and even to Leonardo da Vinci (about 1500). They may be stated as follows:

(1) The magnitude of the frictional force, F, between two solid bodies in contact may be less than, but, in any event, cannot exceed, a definite fraction, μ, of the normal compressive force, N, between the two bodies. Expressed in symbols, the law says $F \leq \mu N$; μ is called the *coefficient of static friction*.

(2) The frictional force, F, attains its greatest possible value, F_l, called the *limiting friction*, when either of the bodies is on the verge of moving. In this case, $F_l = \mu N$. Thus, in a given contact, the frictional

SLIDING FRICTION

force may vary between zero and F_l and, in a state of equilibrium, is just large enough to insure equilibrium.

(3) The limiting value of the frictional force, F_l, is (a) proportional to the normal force, (b) independent of the area of contact, (c) dependent on the nature of the surfaces. This implies that the coefficient of static friction is constant for each combination of materials and surface finish.

(4) The direction of the frictional forces is always such as to oppose the direction in which motion would take place in the absence of friction.

(5) The frictional force after motion begins is called *kinetic* friction. Its relation to the normal force is $F_k = \mu_k N$, where μ_k is called the *coefficient of kinetic friction*. This coefficient is smaller than the coefficient of static friction and, generally, decreases in magnitude as the velocity of sliding increases.

Let us refer to Fig. 66, in which block B, shown resting on the fixed plane A, is subjected to a force of magnitude P sufficient to cause impending motion of the block. All forces shown are acting on B.

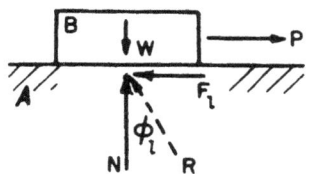

Fig. 66

The resultant reaction of the contact is

$$R = \sqrt{N^2 + F_l^2},$$

where $F_l = \mu N$. The angle between the resultant and the normal is given by

$$\tan \phi_l = \frac{F_l}{N} = \mu.$$

The angle ϕ_l is called the *angle of friction* of the contact. We may note that, as the applied force P changes direction, but remains horizontal, the contact reaction sweeps out a conical surface, called the *cone of friction*.

Consider now a particle resting on a fixed inclined plane (Fig. 67), subjected to no forces except the force of gravity and the contact reaction of the plane. If the particle is at rest, the reaction R must exactly balance the weight and, hence, be in the vertical direction. The equations of equilibrium yield

$$N = W \cos \alpha, \quad F = W \sin \alpha.$$

But $F \leqslant \mu N$, so that $\mu \geqslant \tan \alpha$. Hence, in order for the particle to remain in

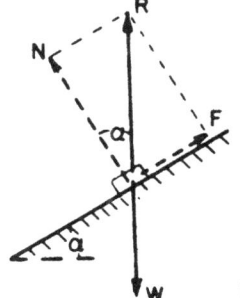

Fig. 67

equilibrium, it is necessary that $\alpha \lessgtr \phi_l$. For this reason, ϕ_l is also called the *angle of repose*.

3. Applications of Coulomb's Laws of Friction

In general, there are two types of frictional problems involving sliding:

(1) Motion impends or occurs on the several surfaces in contact, so that the relations $F_l = \mu N$ or $F_k = \mu_k N$ apply, respectively.

(2) Motion neither impends nor occurs. The frictional force, in this case, is only sufficient to insure equilibrium.

Example 1: A block weighing 10 lbs. rests on a horizontal plane and has a horizontal force of magnitude P applied to it, as shown in Fig. 68. The coefficient of static friction for the contact is $\mu = 0.3$, the coefficient of kinetic friction is $\mu_k = 0.2$. If P takes on, in turn, the values $P = 1, 2, 3, 4$ lbs., determine the force of friction in each case.

The solution proceeds as follows:

$\Sigma F_y = N - 10 = 0$; $N = 10$ lbs.

$F_l = \mu N = 3$ lbs. (applies only when motion impends),

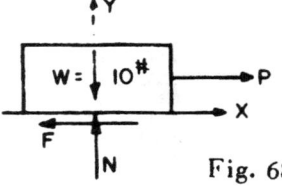

Fig. 68

$F_k = \mu_k N = 2$ lbs. (applies only when motion occurs).

(a) $P = 1$: $\Sigma F_x = 1 - F = 0$; $F = 1$ lb.
(b) $P = 2$: $\Sigma F_x = 2 - F = 0$; $F = 2$ lbs.
(c) $P = 3$: $\Sigma F_x = 3 - F = 0$; $F = 3$ lbs. $= F_l$ (motion impends).
(d) $P = 4$: $\Sigma F_x = 4 - F_k = 2 \neq 0$ (motion occurs).

Example 2: A uniform rod of weight W and length $2l$ rests in a vertical plane with its lower end A on a horizontal floor ($\mu = \mu_1$) and its upper end B against a vertical wall ($\mu = \mu_2$), as shown in Fig. 69. Find the smallest angle θ which the rod can make with the floor before slipping begins.

Since the rod is about to slide, the frictional forces will have attained their respective limiting values. Their directions are shown, in Fig. 69, opposing the direction of im-

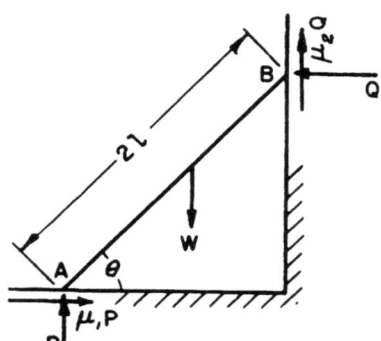

Fig. 69

SLIDING FRICTION

pending motion, in accordance with Coulomb's laws. There remains only to apply the equations expressing equilibrium of the system:

$$\Sigma F_x = \mu_1 P - Q = 0 \atop \Sigma F_y = P + \mu_2 Q - W = 0 \Biggr\} \therefore P = W/(1 + \mu_1\mu_2),\ Q = \mu_1 W/(1 + \mu_1\mu_2).$$

$$\Sigma M_A = Q(2l\sin\theta) + \mu_2 Q(2l\cos\theta) - Wl\cos\theta = 0.$$

Consequently, $\quad \tan\theta = \dfrac{W}{2Q} - \mu_2 = \dfrac{1 - \mu_1\mu_2}{2\mu_1}.$

We note that the required value of θ depends neither on the weight nor on the length of the rod, but only on the coefficients of friction of the two contacts.

Example 3: Find the least value of the horizontal force P applied to the wedge A which will just lift wedge B with its 1000 lb. load, as shown in Fig. 70a. Assume the weight of both wedges to be negligible and take the value of the coefficient of static friction to be $\mu = 0.2$ for all surfaces.

Since sliding is impending on all contacts, the corresponding frictional forces will have attained their maximum possible values, i.e., in all instances, $F = \mu N$. The appropriate free-body diagrams of the wedges B and A are shown in Figs. 70b and 70c.

For equilibrium of B we must have:

$$\Sigma F_x = 0.2 N_C \cos 20 + N_C \sin 20 - N_B = 0,$$

$$\Sigma F_y = N_C \cos 20 - 0.2 N_C \sin 20 - 0.2 N_B - 1000 = 0,$$

from which $N_C = 1306$, $N_B = 692$ lbs.

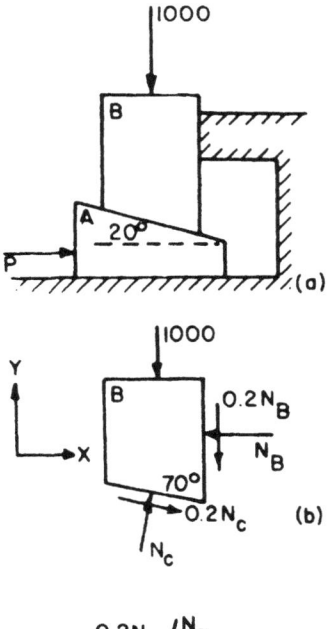
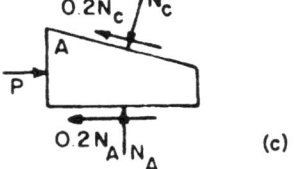

Fig. 70

At first glance it might appear that the above computation is superfluous, since the free-body diagram of A contains three unknown quantities (N_A, N_C, P) and, presumably, we have three equations of equilibrium at our disposal for the determination of these quantities. Such, however, is not the case. For, although we know how to compute the resultant of the normal forces at a given contact, we do not know how these forces are distributed over the surface of the contact; that is, the location of the line of action of the resultant is not known. We may only assume that the forces are distributed in such a way that the condition of moment equilibrium is satisfied identically; the effect of this assumption is to rule out the possibility of tipping. Hence, with reference to Fig. 70c, we may avail ourselves only of the force conditions of equilibrium. These are

$$\Sigma F_y = N_A + 0.2 N_C \sin 20 - N_C \cos 20 = 0,$$
$$\Sigma F_x = P - 0.2 N_C \cos 20 - N_C \sin 20 - 0.2 N_A = 0,$$

and lead to the values $N_A = 1138$, $P = 920$ lbs.

Example 4: A block of weight 100 lbs. and of dimensions as shown in Fig. 71a, rests on an inclined plane ($\mu = 0.2$) and has a horizontal force of magnitude P applied to it. If P is increased gradually, determine whether the block will slide up the plane or tip over.

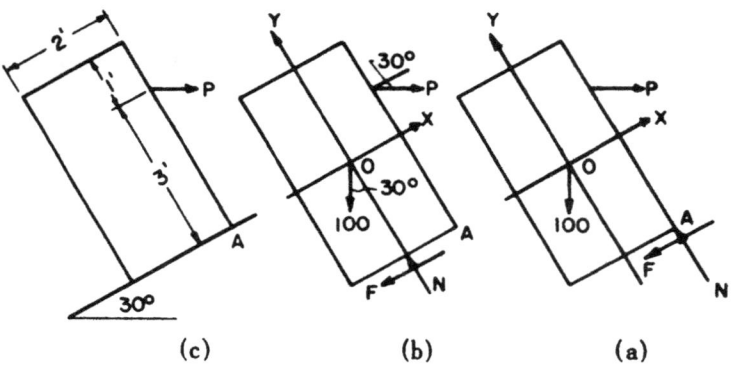

(c) (b) (a) Fig. 71

Assume first that the block slides; the appropriate free-body diagram is shown in Fig. 71b, in which the resultant of the contact pressures, N, is assumed, as in the previous example, to act in such a way as to insure the equilibrium of moments. Then $F = 0.2N$, and

$$\Sigma F_x = P \cos 30 - 100 \sin 30 - 0.2 N = 0,$$
$$\Sigma F_y = N - 100 \cos 30 - P \sin 30 = 0,$$

so that $N = 130.5$ lbs. and $P = 87.8$ lbs.

SLIDING FRICTION

Now assume that the block is about to tip about point A. In this case N acts at A (Fig. 71c) and the value of F will be determined from the conditions of equilibrium. Of these, the equation of moments reads:

$\Sigma M_A = (3)(P \cos 30) - (2)(100 \sin 30) - (1)(100 \cos 30) = 0;\quad P = 71.8$ lbs.

We see that the force P necessary to produce tipping is smaller than the corresponding force needed to cause slipping; hence, the block will tip.

To determine whether sliding accompanies tipping, we compute the force of friction at the contact. Thus,

$\Sigma F_y = N - 100 \cos 30 - P \sin 30 = 0;\quad N = 122.5$ lbs.

$\Sigma F_x = P \cos 30 - 100 \sin 30 - F = 0;\quad F = 12.2$ lbs.

Since the limiting friction $F_l = (0.2)(122.5) = 24.5$ lbs., $F < F_l$, which indicates that pure tipping occurs.

Example 5: Blocks A and B, weighing 200 lbs. and 400 lbs., respectively, are placed on a fixed inclined plane and connected by a flexible cord parallel to the plane, as shown in Fig. 72a. (a) What is the tension in the cord? (b) Does the system move or remain at rest?

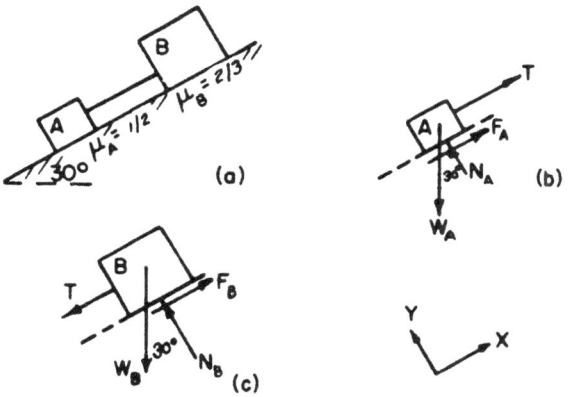

Fig. 72

To compute the tension in the cord we refer to the free-body diagram of block A (Fig. 72b). First of all, we note that, for this block, $\mu_A < \tan 30°$; i.e., the angle of repose of A is smaller than the angle of inclination of the plane (see Section 2), so that A would slide down the plane if it were not restrained by the tension in the cord. This is due to the fact that the component of the weight of A tending to cause sliding exceeds the value of the limiting friction at A. Hence, in the presence of the restraining tension, block A will be on the point of sliding and, therefore, $F_A = \mu_A N_A$. Consequently, the conditions of equilibrium for A are

$$\Sigma F_x = T + \mu_A N_A - W_A \sin 30 = 0,$$
$$\Sigma F_y = N_A - W_A \cos 30 = 0,$$

and their solution is $N_A = 173.2$ lbs., $T = 13.4$ lbs.

We consider, next, the free-body diagram of block B (Fig. 72c) and apply the conditions of equilibrium. The equation

$$\Sigma F_x = F_B - W_B \sin 30 - T = 0$$

leads to the value $F_B = 213.4$ lbs. To find whether the block B moves or remains at rest, we compute the limiting friction at the contact. Thus,

$$\Sigma F_y = N_B - W_B \cos 30 = 0,$$

from which $N_B = 346.4$ lbs., so that the limiting friction at B is $\mu_B N_B = 231$ lbs. Since $F_B < \mu_B N_B$, the block B and, hence, the entire system, remains at rest.

Chapter VII. WORK AND ENERGY METHODS

1. Work

Consider the displacement of a particle, initially at A, along the straight line AB (Fig. 73) under the action of a system of forces. Suppose one of these forces, **F**, to be of *constant* magnitude and direction. We define the *work* of this force as the product of the magnitude of the displacement of its point of application and the magnitude of the component of the force in the direction of the displacement; that is, $W = (F \cos \theta)(AB)$.

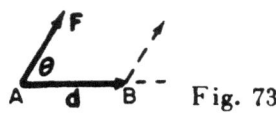

Fig. 73   If we denote by **d** the vector AB, then the work of the force **F** may be expressed as the scalar product $W = \mathbf{F} \cdot \mathbf{d}$. We note that, if $\theta < \pi/2$, $W > 0$, in which case the work is done by the force on the particle; if, on the other hand, $\pi/2 < \theta < \pi$, then $W < 0$, and the work is done by the particle against the action of the force. Of course, for $\theta = \pi/2$, $W = 0$, while for $\theta = 0$, $W = Fd$.

It should be noted that the work done by a system of forces acting simultaneously on a rigid body in the course of displacement of this body is equal to the work done by the resultant of the forces.

Example 1: A box weighing Q lbs. is dragged at constant speed along a rough horizontal floor a distance d ft. by means of a cable. If the angle of inclination, α, of the rope to the floor is constant (Fig. 74), and the coefficient of kinetic friction is μ, compute the work done.

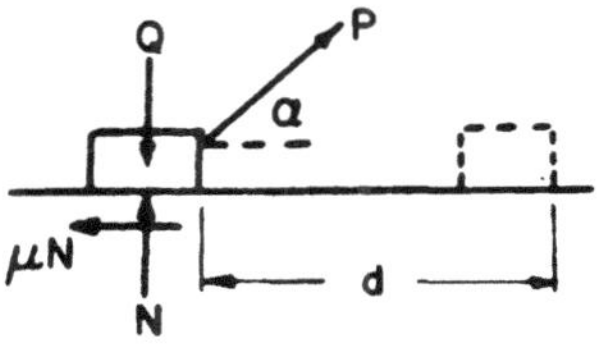

Since the box moves at constant velocity, it is in a state of static equilibrium. Hence,

$$\Sigma F_x = P \cos \alpha - \mu N = 0,$$
$$\Sigma F_y = P \sin \alpha + N - Q = 0,$$

Fig. 74

from which we find $P = \mu Q/(\mu \sin \alpha + \cos \alpha)$. Consequently, the work done by the applied force P is

$$W = (P \cos \alpha) d = \frac{\mu Q d}{1 + \mu \tan \alpha}.$$

Now, assume **F** to vary in magnitude from point to point, always keeping it directed along AB. If we take the coordinate axis Ox along AB, we may describe the variation by stating that $F = F(x)$. This is illustrated by a curve such as is shown in Fig. 75. To compute the work of this force in displacing from A to B, we divide the interval (AB) into small, contiguous, non-overlapping sub-intervals $\Delta x_1, \Delta x_2, \ldots, \Delta x_i, \ldots, \Delta x_n$. In each of these sub-intervals, we may approximate the magnitude of the force by replacing its actual variation by its arithmetic mean in the interval. For example, in the interval Δx_i, let the mean value of $F(x)$ be F_i. Since, now, in each sub-interval the force is constant, we may write an approximate expression for the work of $F(x)$ in (AB):

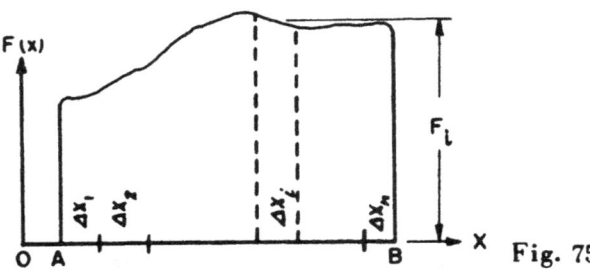

Fig. 75

$$W \approx F_1 \Delta x_1 + F_2 \Delta x_2 + \ldots + F_i \Delta x_i + \ldots + F_n \Delta x_n = \sum_{i=1}^{n} F_i \Delta x_i.$$

This approximation will improve as the number of sub-intervals increases and will yield the exact value in the limit as the number of sub-intervals increases indefinitely. Thus,

$$W = \lim_{\substack{|\Delta x_i| \to 0 \\ n \to \infty}} \sum_{i=1}^{n} F_i \Delta x_i = \int_A^B F(x)\,dx.$$

Example 2: A spring has a modulus of stiffness k whose value is 10 lb./in. (constant). If the initial stretch in the spring is 2 in., find the work done in stretching the spring 4 in.

The modulus of stiffness is defined as the rate of change of the spring force with respect to the spring displacement, i.e., $k = dF/dx$, where x measures the amount of elongation. Thus, in a spring with a constant modulus, the force is proportional to the elongation (linear spring), or $F = kx$. Consequently, in the present case, the work done is

$$W = \int_2^6 (10x)\,dx = 160 \text{ in.lb.}$$

Example 3: A particle P moves along a straight line from A to B under the action of several forces. One of these forces, **F**, whose magnitude is 10 lbs., is always directed toward C (Fig. 76). Find the work done by **F** as P negotiates the distance AB.

The work is given by the expression

$$W = \int_A^B F \cos\theta\, dx;$$

but

$$\cos\theta = \frac{PB}{PC} = \frac{8-x}{\sqrt{(8-x)^2 + (6)^2}},$$

so that

$$W = \int_0^8 \frac{10(8-x)\,dx}{\sqrt{(8-x)^2 + (6)^2}}.$$

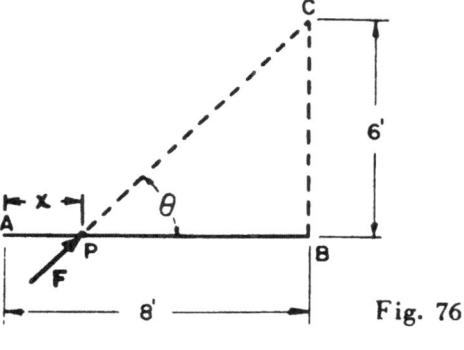

Fig. 76

The substitution $z = 8 - x$ reduces the integral to

$$W = \int_0^8 \frac{10 z\, dz}{\sqrt{z^2 + (6)^2}} = 40 \text{ ft. lbs.}$$

In general, the point of application of the force **F** will traverse a space curve C and the force will vary in direction as well as in magnitude from point to point of the path. Then, with the origin of coordinates at O and **r** the position vector of any point P on the curve with respect to O (Fig. 77), the work done by **F** in displacing from A to B along C is

$$W_{AB} = \int_A^B \mathbf{F} \cdot d\mathbf{r},$$

where $d\mathbf{r} = \mathbf{r}' - \mathbf{r}$ (the change in the position vectors of neighboring points) represents the vector increment measured along the tangent to the path C. Expressed in terms of rectangular components,

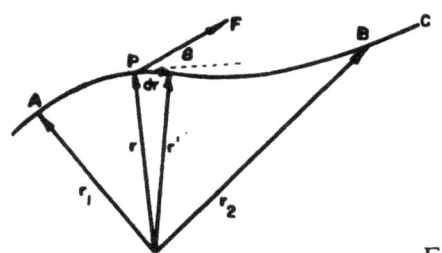

Fig. 77

$$\mathbf{F} = F_x \mathbf{i} + F_y \mathbf{j} + F_z \mathbf{k},$$
$$\mathbf{r} = x\mathbf{i} + y\mathbf{j} + z\mathbf{k},$$
$$d\mathbf{r} = (dx)\mathbf{i} + (dy)\mathbf{j} + (dz)\mathbf{k}.$$

so that

$$W_{AB} = \int_{(x_1,y_1,z_1)}^{(x_2,y_2,z_2)} F_x\,dx + F_y\,dy + F_z\,dz.$$

It is important to note that, since the force **F** varies in magnitude and direction from point to point, each of its components F_x, F_y, F_z will, in general, be a function of all three coordinates x, y, z. The integral expressing W_{AB} differs from the ordinary integral in which the integration takes place over a linear interval, a surface or a volume; in the present case, the integration is performed along a curve, and the integral is called a *line integral*. After F_x, F_y, F_z are specified, the expression for W_{AB} is still indeterminate, since the integral contains information only about the terminal points of the path. But the work done by a force in displacing a particle from one point in space to another will, in general, depend on the particular path which the particle traverses between these terminal points. Thus, the path in any given instance must be specified, and this may be done either by giving the equations of two surfaces, $f(x, y, z) = 0$ and $g(x, y, z) = 0$, whose curve of intersection it represents, or by its parametric form $x = x(t)$, $y = y(t)$, $z = z(t)$. Either formulation completes specification of the problem of integration. In two-dimensional problems, $F_z = 0$ and $z = 0$, so that

$$W_{AB} = \int_{(x_1,y_1)}^{(x_2,y_2)} F_x\,dx + F_y\,dy.$$

Example 4: A particle is displaced in the xy plane from point $A(0,0)$ to point $B(1,1)$ under the action of a force whose components vary in accordance with the expressions $F_x = x - y^2$, $F_y = 2xy$. Compute the work done if the path of the particle is (Fig. 78)

(a) the straight line $y = x$,
(b) the parabola $y = x^2$,
(c) the lines $y = 0$, $0 \leqslant x \leqslant 1$ (AE), and $x = 1$, $0 \leqslant y \leqslant 1$ (EB).

The work is given by the line integral

$$W_{AB} = \int_{(0,0)}^{(1,1)} (x - y^2)\,dx + 2xy\,dy;$$

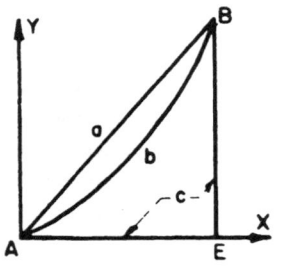

Fig. 78

substitution of the expression for each path reduces the operation to integration with respect to a single variable. Thus,

WORK AND ENERGY METHODS

(a) $y = x$, $dy = dx$;

$$W_{AB} = \int_0^1 [(x - x^2) dx + 2x^2 dx] = \tfrac{5}{6}.$$

(b) $y = x^2$, $dy = 2x\,dx$;

$$W_{AB} = \int_0^1 [(x - x^4) dx + 2x(x^2)(2x\,dx)] = \tfrac{11}{10}.$$

(c) $y = 0$, $0 \leqslant x \leqslant 1$, $dy = 0$; $x = 1$, $0 \leqslant y \leqslant 1$, $dx = 0$;

$$W_{AB} = W_{AE} + W_{EB} = \int_0^1 x\,dx + \int_0^1 2y\,dy = \tfrac{3}{2}.$$

The line integral expressing the work done by a force on a particle may be written in the form

$$W = \int_A^B \mathbf{F} \cdot d\mathbf{r} = \int_A^B \mathbf{F} \cdot \frac{d\mathbf{r}}{dt} dt$$

$$= \int_A^B \mathbf{F} \cdot \mathbf{v}\, dt,$$

where $\mathbf{v}$ is the velocity of the particle with respect to the fixed origin O. The usefulness of this formulation may be seen in the qualitative analysis of the systems shown in Fig. 79. In case (a), a wheel C rolls, without slipping, on a fixed plane A. A frictional force is present at the contact during the rolling process. But the point on the wheel which is, instantaneously, in contact with the plane has zero velocity (the proof of this assertion is relegated to a course on kinematics of rigid bodies). Hence, the frictional force does no work in the course of the motion. This, incidentally, accounts for the relative ease of rolling an object on a rough surface, whereas sliding may necessitate considerable expediture of effort. Theoretically, the work necessary to displace the object by rolling is zero. Actually, some work is required because of the local deformation of the rolling surface at the contact. In case (b), a wheel C rotates about a fixed shaft through its center and is subjected to the action of a brake B.

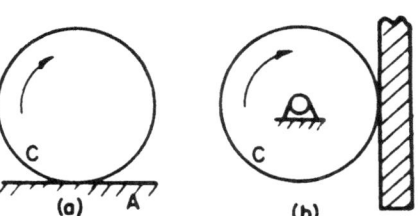

Fig. 79

In this instance, the force of friction at the contact does work since the instantaneous point of contact has a velocity different from zero.

2. The Principle of Virtual Work (for a Free Particle)

Let us consider a free particle (i.e., one devoid of constraints) subjected to a system of forces whose resultant is **R**. Now imagine this particle to receive an arbitrary, very small displacement, $\delta \mathbf{r}$. Such a displacement is called *virtual* to indicate that it is a possible, but not necessarily an actual, displacement.

The work done by a force in the course of a virtual displacement is called *virtual work*; thus, the virtual work of **R** acting through $\delta \mathbf{r}$ is

$$\delta W = \mathbf{R} \cdot \delta \mathbf{r} = X \delta x + Y \delta y + Z \delta z,$$

where X, Y, Z and $\delta x, \delta y, \delta z$ represent the components of **R** and $\delta \mathbf{r}$, respectively, along an arbitrarily chosen system of rectangular coordinate axes.

We recall that, if a particle is in equilibrium, $\mathbf{R} = 0$. Consequently, $\delta W = 0$; that is, a necessary condition for the equilibrium of a particle is that the resultant of all forces acting on it does zero virtual work in the course of an arbitrary, virtual displacement.

Conversely, if the total work of the several forces acting on a particle vanishes for each virtual displacement, then the particle is in equilibrium. This sufficient condition may be proved by noting that $\delta \mathbf{r}$ is completely arbitrary (for a given set of forces). Hence, we may choose, in particular, a virtual displacement for which $\delta y = \delta z = 0$, $\delta x \neq 0$; since $\delta W = 0$ by hypothesis, it follows that $X = 0$. Let us now choose a virtual displacement such that $\delta z = 0$, $\delta y \neq 0$; it having been proved that $X = 0$, we conclude that, in addition, $Y = 0$. Finally, the choice of a displacement for which $\delta z \neq 0$ leads to the result $Z = 0$. Hence, the forces acting on the particle are in equilibrium.

3. Ideal Constraints

It is convenient to distinguish between applied forces and forces arising from various types of constraints and to refer to the former as *active*, to the latter as *reactive* forces. When a given particle is subjected to constraints, any virtual displacement of the particle is no longer completely arbitrary, but must be such that the constraints are not violated. It is desirable to investigate the various possible constraints and to compute the work done by the reactive forces at each constraint in the course of a permissible virtual displacement.

At a smooth contact between a fixed and a movable surface the possible virtual displacement which preserves the contact is a sliding displacement, and, since at such a contact the reaction is normal to the contact surfaces, the reaction does zero work in the course of this displacement. If both surfaces are movable, the reaction on each surface may do work in a contact-preserving displacement; but the reactive forces at the contact are equal and oppositely directed, causing their net virtual work to vanish.

The frictional reaction at a rolling contact between a fixed and a movable surface does zero work in a permissible virtual displacement, as has already been noted in Section 1 of this chapter. The normal component of the reaction at such a contact does zero work because it acts normal to the displacement. At a rolling contact between two movable surfaces, the net work of the pair of contact reactions is zero, in analogy to the situation at the sliding contact of a pair of smooth movable surfaces.

The reaction exerted at any point on a fixed axis of rotation of a rigid body does zero work in the course of a virtual displacement of the body, since all points on the axis of rotation remain fixed.

Finally, it is necessary to consider the action of the constraints which cause any two particles of a rigid body to remain equidistant from each other. The application of a system of active forces to a rigid body will induce internal constraint forces between the various particles of the body which serve to keep the body undeformed and which may be visualized as the reactions P_{mn} and P_{nm} exerted on typical particles m and n by an imaginary rigid rod connecting them (Fig. 80). These internal reactions will be collinear, equal in magnitude, and oppositely directed.

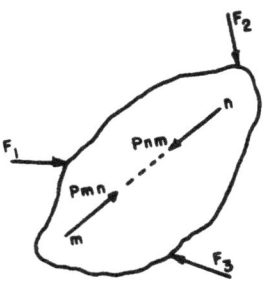

Fig. 80

Now we suppose the system of particles to undergo a virtual displacement compatible with the constraints imposed by the internal reactions; that is, the entire system displaces as a rigid body, causing particles m and n to move to new positions m' and n', respectively (Fig. 81). Any such displacement may be accomplished in two steps: (a) a translation of the body parallel to itself to the intermediate position $m'n_1'$, (b) a rotation, about an axis through m', to the final position $m'n'$. In the course of the translation, the forces P_{mn} and

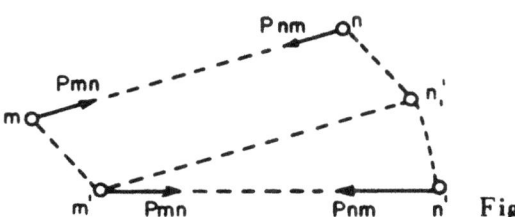

Fig. 81

P_{nm} do non-zero work, but their total net work is zero because the forces are of opposite sign and their displacements are equal $(mm' = nn_1')$. During the rotational portion of the displacement, the force P_{mn} does not displace, since it acts at a point on the fixed axis of rotation, while the force P_{nm} is directed normal to the circular path $n_1' n'$, so that neither force does work. This argument is valid for all pairs of particles of the rigid body and demonstrates that the net virtual work of all internal forces is zero.

In summary, let us tabulate the reactive forces which do zero work in a virtual displacement compatible with the constraints:
(1) the reaction on a movable body in smooth contact with a fixed surface;
(2) the pair of reactions exerted on the two bodies in a smooth contact in which neither body is fixed;
(3) the reaction at a rolling contact between a fixed and a movable body;
(4) the pair of reactions at a rolling contact in which neither body is fixed;
(5) the reaction at any point on a fixed axis of rotation of a rigid body;
(6) the internal forces of constraint in a rigid body.

Constraints at which the reactive forces do zero work in a virtual displacement are called *workless* or *ideal*. A notable exception to ideal constraints is furnished by a rough sliding contact, at which the force of friction causes non-zero work to be done in a sliding displacement.

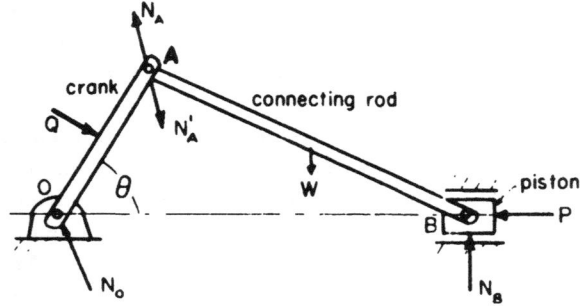

Fig. 82

For an example of a system with ideal constraints let us refer to the *slider-crank* mechanism shown in Fig. 82. It consists of two bars OA and AB, hinged at A; OA (the crank) is constrained to rotate about a fixed axis through O, and OB (the connecting rod) is hinged at B to a piston which is free to slide between smooth walls. The position of the entire system is defined by the single coordinate, θ, so that the only virtual displacement of all particles which does not violate the constraints is the one described by a small (positive or negative) change in θ. Now let us look at the constraint forces and the work done by them when θ is changed by $\delta\theta$. Hinge O is assumed to be frictionless; it remains stationary, so that the reaction N_O does no work. Reactions N_A and N_A' are equal in magnitude and oppositely directed; although each of them does non-zero work in the course of the stipulated virtual displacement, the net work done by them is zero. The piston B slides in smooth contact so that the reaction N_B remains normal to the displacement and, hence, does no work. Finally, the internal forces between particles which keep the various members of the mechanism rigid do zero net work in the course of the displacement of the system. Thus, the net work of all forces of constraint is zero.

4. The Principle of Virtual Work (for Systems)

Let us now focus our attention on a system of particles retained in position by means of ideal constraints and subjected to applied forces. If the system is in equilibrium, each particle must be in equilibrium. Consequently, the work of all applied and reactive forces acting on any particle in a virtual displacement will be zero. If $\mathbf{P}_i$ and $\mathbf{N}_i$ denote, respectively, the resultant active and reactive forces on the ith particle and $\delta \mathbf{r}_i$ the virtual displacement, then $(\mathbf{P}_i + \mathbf{N}_i) \cdot \delta \mathbf{r}_i = 0$. Since a similar equation holds for each of the n particles of the system, we find, by adding all the equations, $\sum_{i=1}^{n} (\mathbf{P}_i + \mathbf{N}_i) \cdot \delta \mathbf{r}_i = 0$. But $\sum_{i=1}^{n} \mathbf{N}_i \cdot \delta \mathbf{r}_i = 0$ identically for a system with ideal constraints, so that

$$\sum_{i=1}^{n} \mathbf{P}_i \cdot \delta \mathbf{r}_i = 0.$$

We see that the work of all *active* forces is zero in an arbitrary virtual displacement; this represents a necessary condition for equilibrium of the system.

We may prove sufficiency of this condition by assuming that the equation holds but that equilibrium does not exist. Therefore, each particle will begin to move, without violating its constraints, in the direction of the resultant of all forces acting on it. Consequently, non-zero work will be produced. But the work of the constraints is zero, which implies that the work of the active forces must be different from zero. Since the hypothesis is contradicted, the resultant of the active forces must be zero and equilibrium must obtain.

We have arrived at an alternative (but completely equivalent) formulation of the conditions of equilibrium, expressed by the

Principle of Virtual Work: A necessary and sufficient condition for the existence of equilibrium of a system of particles with ideal constraints is the vanishing of the work done by the active forces in an arbitrary, infinitesimal, virtual displacement compatible with the constraints of the system.

The principle of virtual work, coupled with the expression for a general displacement of a rigid body (the latter derivable from kinematics), may be used to demonstrate the sufficiency of the conditions of vanishing force and moment which were shown, in Chapter II, Section 4, to be necessary for the existence of equilibrium of a system of particles. The proof is omitted here.

Example 1: Blocks A and B, in contact with the smooth, inclined planes D and E, respectively, are connected by means of a string which passes over a smooth pulley C, as illustrated in Fig. 83. What must be the relation between the weight of A (P) and the weight of B (Q) in order for the system to be in equilibrium?

We note that a small displacement, δs, of A up the plane D is a permissible virtual displacement of the system. At the same time, B will descend along E by the same amount. The only active forces being the two forces of gravity, the principle of virtual work yields the relation

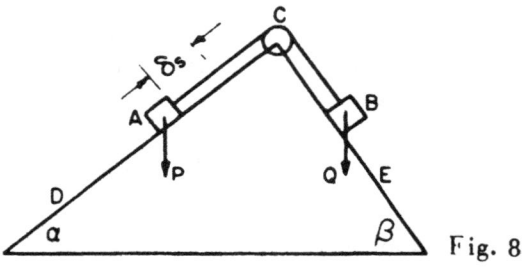

Fig. 83

$$\delta W = -P \sin \alpha \, \delta s + Q \sin \beta \, \delta s = 0,$$

from which we find $P/Q = \sin \beta / \sin \alpha$.

Example 2: A block of weight Q is maintained in equilibrium on a smooth plane of inclination α by a force P which acts (a) horizontally, (b) upward parallel to the plane, as shown in Figs. 84a, b. Compute the magnitude of the force in each case.

In each case we may imagine the block to undergo a virtual displacement δs upward and parallel to the plane. Therefore, in case (a), $\delta W = (P \cos \alpha) \delta s - (Q \sin \alpha) \delta s = 0$, or $P = Q \tan \alpha$. In case (b), $\delta W = P \delta s - (Q \sin \alpha) \delta s = 0$, or $P = Q \sin \alpha$.

WORK AND ENERGY METHODS

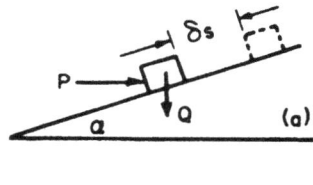

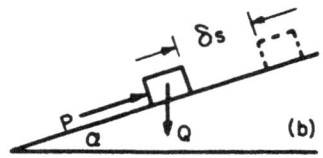

Fig. 84

Example 3: A thin wire is shaped to form a circular hoop fixed in the vertical plane. A small bead B of weight Q is fastened to the hoop in such a way that it can slide freely along the wire. A string is attached to the bead and passed over a pulley A located at the top of the vertical diameter, with a weight P freely suspended from the other end of the string (Fig. 85). Find the positions of equilibrium, measured by the angle θ the string makes with the vertical.

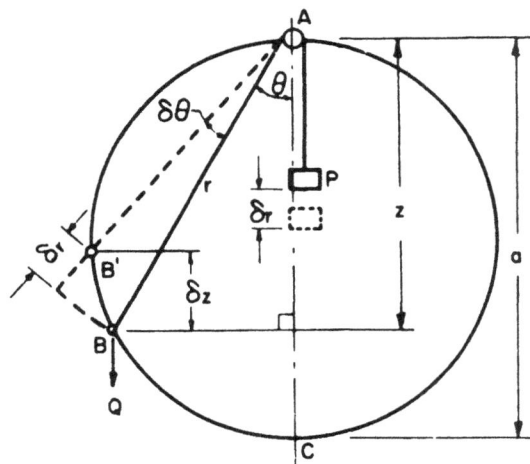

Fig. 85

Let r and z denote the radial distance and the vertical distance, respectively, which separate the bead from the pulley. Thus, in a virtual displacement of the bead upward along the circular arc to position B', the bead will have displaced through a vertical distance δz, the weight P downward through a distance δr, and the angle between the string and the vertical will have increased by an amount $\delta\theta$. Applying the principle of virtual work, we find

$$\delta W = -Q\delta z + P\delta r = 0.$$

But δz and δr are not independent, since they both vary with $\delta\theta$. We note that, since $\angle CBA = 90°$, $r = a\cos\theta$ and $z = r\cos\theta = a\cos^2\theta$; hence, on differentiating, we get

$$\delta r = -a\sin\theta\,\delta\theta, \quad \delta z = -2a\sin\theta\cos\theta\,\delta\theta.$$

Thus, $(P - 2Q\cos\theta)\sin\theta\,\delta\theta = 0$. Since $\delta\theta$ is arbitrary, we find two possible solutions to our problem:

    (a) $\sin\theta = 0$, i.e., $\theta = 0$ (particle at point C);
    (b) $\cos\theta = P/2Q$ (for a real solution, $P < 2Q$).

Thus far we have applied the principle of virtual work to the calculation of active forces, the advantage of the method in each case deriving from the fact that the forces of constraint did not need to be considered. However, it is possible to employ the method of virtual work to compute reactive forces. The appropriate constraint is simply removed and replaced by its force of reaction (which has, by this procedure, been transformed into an active force). At the same time, removal of the constraint creates for the system a permissible, but hitherto forbidden, virtual displacement in the course of which the reactive force performs work. The procedure is best illustrated by examples.

Example 4: A uniform beam AC, l ft. long and weighing q lb./ft., is hinged at the right end A and supported by a roller at B, a distance b ft. from A. If a force P is applied at a point d ft. to the left of A (Fig. 86a), compute the reaction on the beam due to the roller at B.

There is no virtual displacement which the system shown can undergo without violating any of the constraints. However, if we remove the constraint at B (the roller) and replace it by its reactive force R_B,

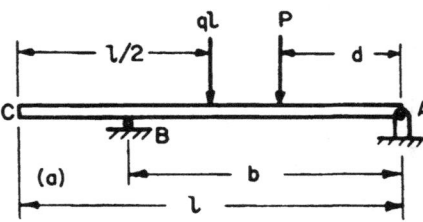

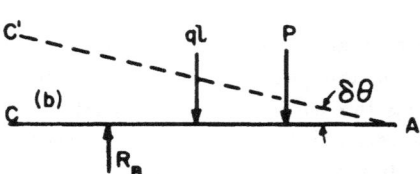

Fig. 86

the system may then be displaced by a virtual rotation of amount $\delta\theta$ about A to a position AC' (Fig. 87b). Hence, for equilibrium to obtain,

$$\delta W = R_B(b\,\delta\theta) - ql\left(\frac{l}{2}\delta\theta\right) - P(d\,\delta\theta) = 0$$

so that

$$R_B = (Pd + \tfrac{1}{2}ql^2)/b.$$

In the preceding computation the forces were assumed normal to the axis of the beam. If, alternatively, the forces had been taken to remain vertical in the course of the rotation about A, the individual expressions for virtual work would have involved $\sin\delta\theta$ instead of $\delta\theta$, with the identical result.

Example 5: In the latticework shown in Fig. 87a, compute the reaction in the crossbar BC due to the load L applied at D, the weight of the members being assumed negligible.

Removal of the constraint (bar BC) causes introduction of reactive forces S (assumed compressive) acting on hinges B and C. The system may now be subjected to a virtual displacement in which BA and CA each rotate about the hinge at A by a virtual angle $\delta\theta$ but in opposite directions (Fig. 87b). The principle of virtual work yields

$$\delta W = 2S\,\delta x_1 - L\,\delta x_2 = 0.$$

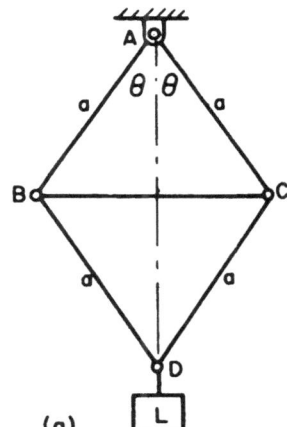

We now seek a relation to express δx_1 and δx_2 in terms of the independent angle of rotation $\delta\theta$. From Fig. 87b we find

$$\delta x_1 = a\sin(\theta + \delta\theta) - a\sin\theta$$
$$\approx a\cos\theta\,\delta\theta,$$

$$\delta x_2 = 2a\cos\theta - 2a\cos(\theta + \delta\theta)$$
$$\approx 2a\sin\theta\,\delta\theta,$$

the approximations at the extreme right being valid because $\delta\theta \ll 1$. Hence,

$$S = L\tan\theta \text{ (compressive)}.$$

(a)

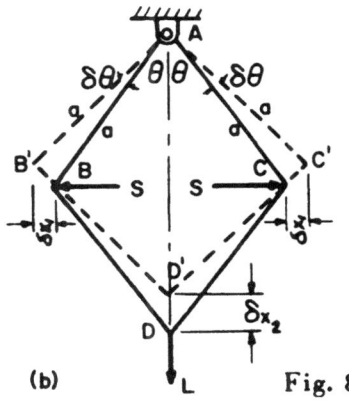

(b)

Fig. 87

Example 6: Referring to the truss shown in Fig. 88a, compute the reactions in members AD and DF.

To find the reaction in AD, we remove the constraint and thus permit the virtual displacement in which AC and BD rotate through an angle $\delta\theta$ about A and B, respectively (Fig. 88b). The portion of the truss above CD, being rigid, remains vertical after having displaced horizontally by an amount $\delta x \approx (h\,\delta\theta)\cos\delta\theta \approx h\,\delta\theta$ and vertically by an amount $(h\,\delta\theta)\sin\delta\theta$. The latter displacement is neglected since, for $\delta\theta \ll 1$, it represents an infinitesimal of order higher than δx. Hence, on noting that the force R at A does zero work, we find

$$\delta W = P_1\,\delta x + P_2\,\delta x + P_3\,\delta x - (R\cos\alpha)\delta x = 0,$$

so that

$$R = (P_1 + P_2 + P_3)/\cos\alpha$$
(tension).

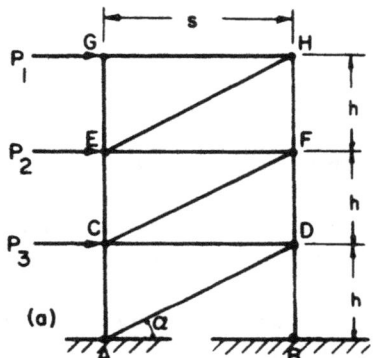

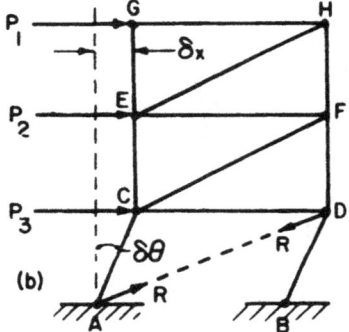

To compute the force in DF, we remove that member, enabling the structure to undergo the virtual rotation by an amount $\delta\phi$ about C (Fig. 88c). Once again the vertical displacements of joints G and E are infinitesimals of higher order when compared with the horizontal displacements of these joints. Joint F displaces, as a result of the rotation of CF by $\delta\phi$ about C, in the horizontal direction by an amount $s\tan\alpha\;\delta\phi$ and in the vertical direction by an amount $s\delta\phi$;

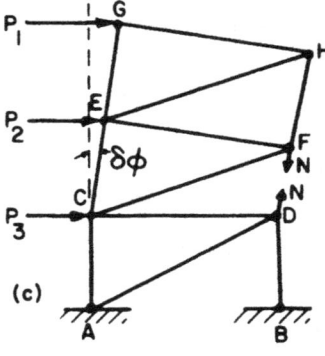

Fig. 88

hence, to quantities of order $\delta\phi$,
$$\delta W = P_1 (2h\,\delta\phi) + P_2 (h\,\delta\phi) + N(s\,\delta\phi) = 0,$$
from which we find
$$N = -\frac{h}{s}(2P_1 + P_2) \quad \text{(compression)}.$$

5. Potential Energy

We refer to a set of positions of all particles of a system as the *configuration* of the system. As has been shown in Section 1 of this chapter, the work produced by a set of forces acting on a particle or a system of particles depends, in general, on the path along which the particles displace. Let us investigate the conditions under which the work done by a set of forces acting on a system of particles is the same for all paths which the system may take in passing from a configuration A to any other configuration B. What we mean is that the work of the forces is to be a function only of the terminal points of the path (mathematically, of the limits of integration), i.e.,

$$W = \int_A^B \mathbf{F} \cdot d\mathbf{r} = \Phi(B) - \Phi(A).$$

But, in order for this equation to hold, the integrand must be an exact differential, i.e.,
$$dW = \mathbf{F} \cdot d\mathbf{r} = d\Phi,$$
or, X, Y, Z being components of $\mathbf{F}$,
$$X dx + Y dy + Z dz = \frac{\partial \Phi}{\partial x} dx + \frac{\partial \Phi}{\partial y} dy + \frac{\partial \Phi}{\partial z} dz.$$

Since the increments dx, dy, dz are arbitrary and independent, we find
$$X = \frac{\partial \Phi}{\partial x}, \quad Y = \frac{\partial \Phi}{\partial y}, \quad Z = \frac{\partial \Phi}{\partial z}.$$

Thus, the condition for the work of a force to be independent of path is that the rectangular components of the force be expressible in terms of the respective derivatives of a single function $\Phi(x, y, z)$. The force field (a collective term which denotes the variation of force from point to point in space) is said to be *conservative* and Φ is called a *force potential*.

Confining our attention to conservative forces, let us choose one configuration of our system of particles to be a *standard* configuration and denote this by A_0. We define a function $V(A)$ as the work done in displacing the system from an arbitrary configuration A to the standard configuration A_o:

$$V(A) = W_{AA_o} = \int_A^{A_o} \mathbf{F} \cdot d\mathbf{r} = \Phi(A_o) - \Phi(A).$$

Thus, $V(x,y,z) = -\Phi(x,y,z) + \text{constant}$. $V(A)$ is called the *potential energy* of the system in the configuration A with respect to the standard configuration A_o and is defined to within an additive constant. The force field is related to the potential energy by means of the expressions

$$X = -\frac{\partial V}{\partial x}, \quad Y = -\frac{\partial V}{\partial y}, \quad Z = -\frac{\partial V}{\partial z}.$$

In two dimensional systems, $V = V(x,y)$ and

$$X = -\frac{\partial V}{\partial x}, \quad Y = -\frac{\partial V}{\partial y}.$$

The simplest example of a conservative force field is furnished by the gravitational field acting on a particle on the surface of the earth. Referred to a system of rectangular coordinates whose z-axis is vertical and directed upward (Fig. 89), this force field has components

$$X = Y = 0, \quad Z = -W\mathbf{k},$$

where W is the weight of the particle. Hence, the potential energy function is $V = Wz + \text{constant}$. The arbitrary constant will be specified by the choice of a standard configuration. In the notation of Fig. 89, the potential energy at A with respect to A_o is given by

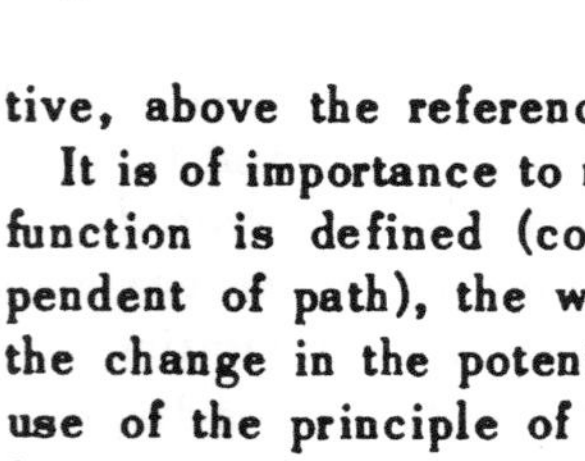

$$V = \int_A^{A_o} \mathbf{F} \cdot d\mathbf{r} = \int_z^{z_o}(-W\,dz) = W(z - z_o).$$

Fig. 89

The potential energy of a gravitational field on the surface of the earth is given by the product of the weight and the vertical height, positive or negative, above the reference position.

It is of importance to note that in a system for which a potential energy function is defined (conservative force field—work of forces is independent of path), the work done in a virtual displacement is related to the change in the potential energy, i.e., $dW = -dV$. Hence, by making use of the principle of virtual work, we find that a necessary and sufficient condition for a conservative system to be in equilibrium is the vanishing of the change in potential energy in any infinitesimal virtual displacement. This, in turn, implies that the potential energy has a stationary value in a position of equilibrium. Here, then, is another criterion for determining static equilibrium.

6. Stability

A system is said to be in *stable* equilibrium if, following a displacement from its configuration by a small force applied for a short time, it returns to its original position as soon as the disturbing force is removed. Such a system is illustrated in Fig. 90a. The potential energy attains a *minimum* value in such a position of equilibrium.

A system is said to be in *unstable* equilibrium if, as a result of the momentary action of a small disturbing force, motion of the system occurs away from its original position. A typical system is shown in Fig. 90b. The potential energy is at a *maximum* in this case.

A system is said to be in *neutral* equilibrium if, subsequent to a small, momentary disturbance, it remains in equilibrium in any new neighboring configuration. An illustration of this condition is given in Fig. 90c. In such an instance the potential energy function is at a *minimax* (in three dimensions, the surface representing the potential energy has a saddle point; in two dimensions, the curve representing the potential energy has a point of inflection).

(a) stable (b) unstable (c) neutral   Fig. 90

Example 1: A thin, rectangular plate $ABCD$, of sides a and b, is suspended from a smooth vertical wall OE by a string OA, of length b, attached to one corner of the plate. The plate is in equilibrium with a second corner in contact with the wall, as shown in Fig. 91. Find the angle θ which the string makes with the wall and determine the type of equilibrium.

If we let the reference height of zero potential energy be on a horizontal line through point O, then the potential energy of the plate is

$$V = Wy,$$

where y represents the vertical distance of F above Ox; its absolute value is given by the sum of the vertical projections of OA, AG and GF. Thus,

$$V = W\left[-\left(b\cos\theta + \frac{b}{2}\cos\theta + \frac{a}{2}\sin\theta\right)\right] = -\frac{W}{2}(3b\cos\theta + a\sin\theta).$$

For equilibrium,

$$\frac{dV}{d\theta} = \frac{W}{2}(3b\sin\theta - a\cos\theta) = 0,$$

so that

$$\theta = \arctan(a/3b).$$

Stability is determined by the sign of the second derivative at the position of equilibrium:

$$\left.\frac{d^2V}{d\theta^2}\right|_{\theta_{eq.}} = \frac{W}{2}(3b\cos\theta + a\sin\theta)\bigg|_{\theta_{eq.}} = \frac{W}{2}\sqrt{a^2 + 9b^2} > 0.$$

Since the second derivative is positive, the potential energy is at a minimum and the equilibrium is stable.

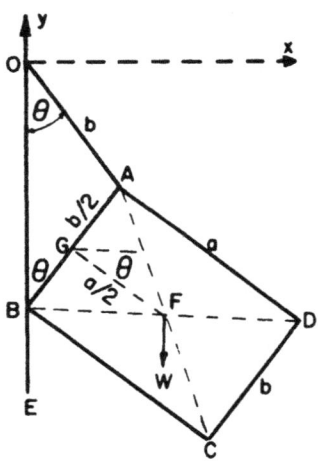

Fig. 91

Example 2: A rod AB, of length l and weight W, is supported by a fixed smooth curve OC and rests against a smooth vertical wall OB, as shown in Fig. 92. Find the equation of the curve OC, referred to rectangular coordinate axes Ox and Oy, for which the rod will be in equilibrium in any position as defined by the angle θ ($< 90°$).

Since the rod is to be in equilibrium for any (i.e., every) value of θ, the potential energy must be constant. Hence, the curve OC must be such that the vertical height of the center of the rod, D, above Ox remains the same. In the position shown in Fig. 92, this height is

$y + (l/2)\cos\theta$; if the rod were in the position in which A would coincide with O, the height would be $l/2$. Therefore,

$$y + (l/2)\cos\theta = (l/2);$$

but

$$\cos\theta = \sqrt{l^2 - x^2}/l,$$

so that

$$2y = l - \sqrt{l^2 - x^2}.$$

The curve is an ellipse.

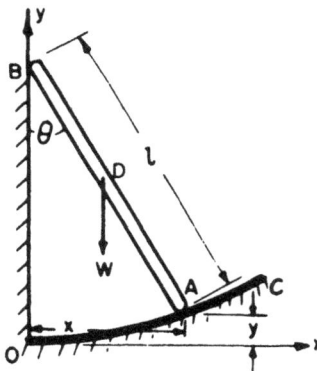

Fig. 92

Example 3: A cylindrical container rests on top of a fixed sphere, as shown in Fig. 93a. What must be the relation between the radius, r, of the sphere and the height, c, of the center of gravity of the container above its base for stable equilibrium to exist?

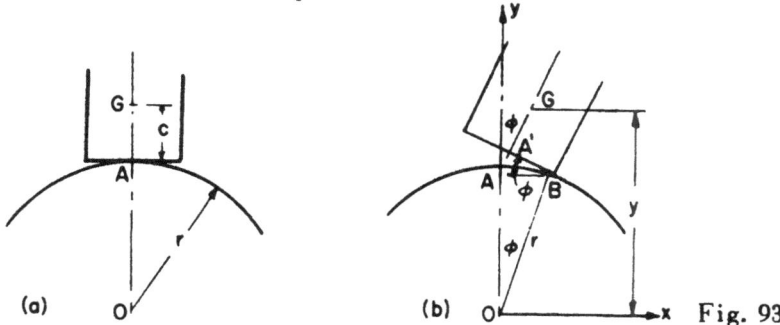

Fig. 93

In each of the two previous examples the system was in an arbitrary configuration, so that the general expression for the potential energy could be written immediately. However, in the present problem, the system is in a very special configuration (i.e., in equilibrium); hence, in order that we may write the expression for the potential energy in an arbitrary position, the system must be displaced from equilibrium without violating the constraints. This is accomplished by rolling the container on the sphere to the position shown in Fig. 93b. Now the potential energy of the container can be computed. With the horizontal reference line taken through the center of the fixed sphere, $V = Wy$, where W is the weight of the container and y the height of its center of gravity, G, above Ox. The latter may be obtained by adding the vertical projections of the lines OB, BA' and $A'G$. It will be noted that A' is the point on the container which, in a position of equilibrium, is in contact with point A on the sphere. Since the container, in passing from the configuration of Fig. 93a to that of Fig. 93b, was assumed to have rolled without slipping, $BA' = BA = r\phi$. Thus,

$$y = r \cos \phi + r\phi \sin \phi + c \cos \phi,$$

and

$$\frac{dV}{d\phi} = W(r\phi \cos \phi - c \sin \phi).$$

As expected, $\dfrac{dV}{d\phi} = 0$ at $\phi = 0$ (position of equilibrium). To examine stability, we compute

$$\frac{d^2V}{d\phi^2} = W[r(\cos \phi - \phi \sin \phi) - c \cos \phi].$$

At the position of equilibrium ($\phi = 0$),

$$\left.\frac{d^2V}{d\phi^2}\right|_{\phi=0} = W(r - c).$$

If equilibrium is to be stable, this quantity must be positive; hence, $c < r$.

Note should be taken of the fact that in each of the three preceding problems the reference line of zero potential energy was taken through a point *fixed* in space.

Example 4: It is instructive to alter the problem of Example 5, p. 77, to a form in which it is suitable for the potential energy approach. The latticework of Fig. 87a is shown in Fig. 94, but with the web member BC omitted. In its place are two horizontal forces T, applied at the joints B and C. Required is the angle θ at equilibrium.

WORK AND ENERGY METHODS

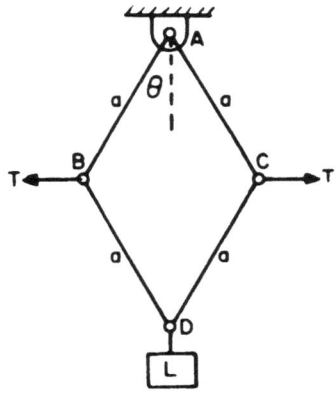

Fig. 94

We are no longer dealing solely with vertical forces, so that the horizontal line of reference employed heretofore must now be replaced by a standard configuration. The standard may be chosen to be the position for which the latticework is folded along the vertical direction, i.e., for which $\theta = 0$. Recalling that, by definition, the potential energy of an arbitrary, with respect to a standard, configuration is the work done by the forces in their passage from the arbitrary to the standard position, we find

$$V = -2T(a \sin \theta) + L(2a - 2a \cos \theta)$$
$$= 2a[-T \sin \theta + L(1 - \cos \theta)].$$

For equilibrium,

$$\frac{dV}{d\theta} = 2a(-T \cos \theta + L \sin \theta) = 0,$$

so that

$$\theta = \text{arc tan } (T/L).$$

If each of the bars in Fig. 94 is of weight w, the effect on the equilibrium configuration may be evaluated by taking account of the potential energy of the weights. The additional terms are

$$2w\left(\frac{a}{2} - \frac{a}{2} \cos \theta\right) + 2w\left(\frac{3a}{2} - \frac{3a}{2} \cos \theta\right) = 4wa(1 - \cos \theta),$$

so that the total potential energy of the system is

$$V = 2a[-T \sin \theta + (L + 2w)(1 - \cos \theta)].$$

Hence, the condition $dV/d\theta = 0$ yields

$$\theta = \text{arc tan } T/(L + 2w).$$

Chapter VIII. EQUILIBRIUM OF SIMPLE SPATIAL SYSTEMS

In Section 4 of Chapter II we found that the conditions necessary to insure equilibrium of a system require

$$\mathbf{F} = 0, \quad \mathbf{M} = 0,$$

that is, the vanishing of the resultant of all external forces acting on the system and of the resultant of the moments of all external forces about an arbitrary point in space. Whereas, applied to planar systems, these equations are equivalent to three scalar equations, in problems of systems in space they are equivalent to six scalar equations.

It will be sufficient to illustrate the procedure involved in problems of equilibrium of spatial systems by means of examples involving, in turn, concurrent, parallel and general force systems.

Example 1 (Concurrent Forces): A bracket made up of six bars joined together at A and B and joined to a vertical wall at C, D, E and F, carries a vertical load P at the joint B, as shown in Fig. 95. The joints A, B, E, F lie in one horizontal plane and joints C and D are located vertically below E and F, respectively. Compute the reaction in each bar of the structure.

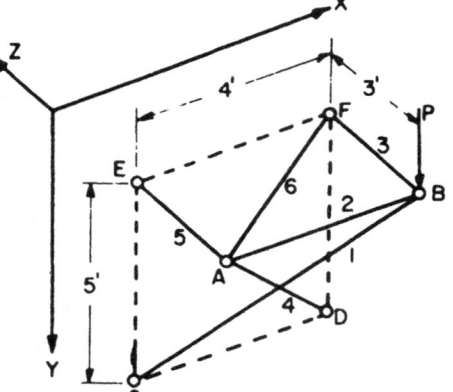

Fig. 95

We consider, first, the free-body diagram of joint B; having assumed each of the reactive forces to be tensile, we may write the equation of the equilibrium of forces in the form:

$$P\mathbf{j} + T_1\left(-\frac{4}{5\sqrt{2}}\mathbf{i} + \frac{5}{5\sqrt{2}}\mathbf{j} + \frac{3}{5\sqrt{2}}\mathbf{k}\right) - T_2\mathbf{i} + T_3\mathbf{k} = 0.$$

On setting the coefficient of each unit vector equal to zero, we obtain the scalar equations

At Joint B

$$\Sigma F_x = -\frac{4T_1}{5\sqrt{2}} - T_2 = 0,$$

$$\Sigma F_y = P + \frac{5T_1}{5\sqrt{2}} = 0,$$

$$\Sigma F_z = \frac{3T_1}{5\sqrt{2}} + T_3 = 0,$$

from which we find

$T_1 = -P\sqrt{2}$ (compression),
$T_2 = 0.8P$ (tension),
$T_3 = 0.6P$ (tension).

Now, let us analyze the equilibrium of joint A. The (vector) force equation in this case is

$$T_2\mathbf{i} + T_4\left(\frac{4}{5\sqrt{2}}\mathbf{i} + \frac{5}{5\sqrt{2}}\mathbf{j} + \frac{3}{5\sqrt{2}}\mathbf{k}\right) + T_5\mathbf{k} + T_6\left(\frac{4}{5}\mathbf{i} + \frac{3}{5}\mathbf{k}\right) = 0,$$

from which we obtain the scalar equations

At Joint A

$$\Sigma F_x = T_2 + \frac{4T_4}{5\sqrt{2}} + \frac{4T_6}{5} = 0,$$

$$\Sigma F_y = \frac{5T_4}{5\sqrt{2}} = 0,$$

$$\Sigma F_z = \frac{3T_4}{5\sqrt{2}} + T_5 + \frac{3T_6}{5} = 0.$$

Hence, $T_4 = 0$, $T_5 = 0.6P$ (tension), $T_6 = -P$ (compression).

Example 2 (Parallel Forces): A uniform rectangular plate $ABCD$, of width a, length b, and weight W, is supported in a horizontal position by three vertical bars hinged as shown in Fig. 96. Determine the force in each supporting bar.

If we assume each of the reactive forces to be tensile, we find the equation of equilibrium of forces

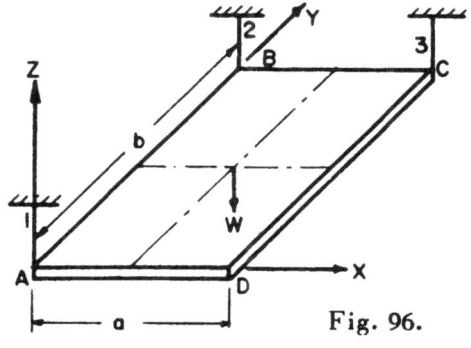

Fig. 96.

i.e.,
$$T_1\mathbf{k} + T_2\mathbf{k} + T_3\mathbf{k} - W\mathbf{k} = 0,$$
$$\Sigma F_x = \Sigma F_y = 0,$$
$$\Sigma F_z = T_1 + T_2 + T_3 - W = 0.$$

The equation of equilibrium of moments is

$$\Sigma M_A = b\mathbf{j} \times T_2\mathbf{k} + (b\mathbf{j} + a\mathbf{i}) \times T_3\mathbf{k} + \left(\frac{b}{2}\mathbf{j} + \frac{a}{2}\mathbf{i}\right) \times (-W\mathbf{k}) = 0,$$

or, expressed in terms of scalar components,

$$\Sigma M_x = b\left(T_2 + T_3 - \frac{W}{2}\right) = 0,$$
$$\Sigma M_y = a\left(-T_3 + \frac{W}{2}\right) = 0,$$
$$\Sigma M_z = 0.$$

Hence, $T_1 = \dfrac{W}{2}$, $T_2 = 0$, $T_3 = \dfrac{W}{2}$.

Example 3 (General Forces): A uniform bar AB, of length d and weight W, is supported at its lower end, A, in the corner of a box of width b and height c. The upper end, B, of the bar is in contact with a smooth wall of the box and is prevented from sliding down by a string BC, of length l, attached to the upper corner, C, of the box, as shown in Fig. 97. Compute the tension in the string and the compressive reaction against the box at A and B.

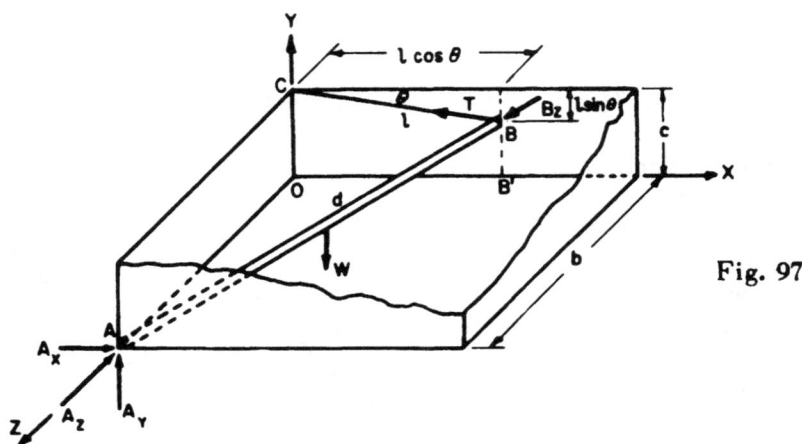

Fig. 97

EQUILIBRIUM OF SIMPLE SPATIAL SYSTEMS

The condition of force equilibrium reads

$$\Sigma \mathbf{F} = (A_x \mathbf{i} + A_y \mathbf{j} - A_z \mathbf{k}) + B_z \mathbf{k} + T(-\cos\theta \mathbf{i} + \sin\theta \mathbf{j}) - W \mathbf{j} = 0,$$

the condition for moment equilibrium about A is

$$\Sigma \mathbf{M}_A = (c\mathbf{j} - b\mathbf{k}) \times T(-\cos\theta \mathbf{i} + \sin\theta \mathbf{j})$$
$$+ (c\mathbf{j} - b\mathbf{k} + l\cos\theta \mathbf{i} - l\sin\theta \mathbf{j}) \times B_z \mathbf{k}$$
$$+ \tfrac{1}{2}(c\mathbf{j} - b\mathbf{k} + l\cos\theta \mathbf{i} - l\sin\theta \mathbf{j}) \times (-W\mathbf{j}) = 0.$$

In terms of equations relating scalar components, these become
(1) $\Sigma F_x = A_x - T\cos\theta = 0$,
(2) $\Sigma F_y = A_y + T\sin\theta - W = 0$,
(3) $\Sigma F_z = -A_z + B_z = 0$,
(4) $\Sigma M_x = bT\sin\theta + (c - l\sin\theta)B_z - \tfrac{1}{2}bW = 0$,
(5) $\Sigma M_y = bT\cos\theta - lB_z\cos\theta = 0$,
(6) $\Sigma M_z = cT\cos\theta - \tfrac{1}{2}lW\cos\theta = 0$.

From (6), $T = Wl/2c$;
from (1), $A_x = (Wl/2c)\cos\theta$;
from (2), $A_y = W[1 - (l/2c)\sin\theta]$;
from (3) and (5), $A_z = B_z = Wb/2c$.

Equation (4) is then satisfied identically. The angle θ may be expressed in terms of the dimensions given in the statement of the problem by noting that AB', which is the projection of AB on the xz plane, is given by

$$AB' = [d^2 - (c - l\sin\theta)^2]^{1/2} \quad \text{(from rt. } \triangle ABB')$$
$$= [b^2 + (l\cos\theta)^2]^{1/2} \quad \text{(from rt. } \triangle AOB').$$

Hence,
$$\sin\theta = \frac{b^2 + c^2 + l^2 - d^2}{2cl}.$$

Appendix. CENTER OF GRAVITY

1. Static Equivalence of Parallel Forces

Let us consider a system of forces having parallel lines of action and determine the location of the resultant. Since the forces are parallel they will have a common unit vector, n_0; hence, a typical member of the system may be expressed by $F_i = F_i n_0$.

To find the single force, R, statically equivalent to our system, we note that, in accordance with our definition of static equivalence (p. 41), the single force must be equal to the sum of the forces of the system, i.e.,

$$R = \sum_{i=1}^{n} F_i = \left(\sum_{i=1}^{n} F_i \right) n_0.$$

Moreover, the moment of R with respect to an arbitrary point, A, must be equal to the sum of the moments of the individual forces of the system about A. Thus, if r_i denotes the position vector of F_i and $\bar{r}$ the position vector of R, each with respect to A, then the condition on the moments may be written

$$M_A = \sum_{i=1}^{n} r_i \times F_i = \bar{r} \times R.$$

If we now take account of the expressions for R and F_i, we find

$$\sum_{i=1}^{n} r_i \times (F_i n_0) = \bar{r} \times \left(\sum_{i=1}^{n} F_i \right) n_0,$$

or

$$\left(\sum_{i=1}^{n} F_i r_i - \bar{r} \sum_{i=1}^{n} F_i \right) \times n_0 = 0.$$

Since $n_0 \neq 0$ and the vector representing the quantity in parentheses is not, in general, parallel to n_0, the only way in which this equation can be satisfied is if the quantity in parentheses is identically zero. Accordingly, we obtain the following expression which determines the *line of action* of **R**:

$$\bar{r} = \sum_{i=1}^{n} F_i r_i \bigg/ \sum_{i=1}^{n} F_i.$$

In the event that F_i represents a system of bound forces, i.e., forces whose respective points of application, P_i, are specified (in our case, by the position vectors r_i), the formula for $\bar{r}$ yields the position vector of a point, which we may call the *center* of the system of parallel forces.

2. Center of Gravity

We now wish to apply the results of the preceding investigation to the particular case of gravitational forces exerted on a given body at the surface of the earth.

Referring to Fig. 98, we imagine the body subdivided into a large number of elements, each of volume $\Delta_i V$ and density ρ_i, and hence, of weight $F_i = \rho_i \Delta_i V$. Accordingly, the center of these forces, measured with respect to an arbitrary point A, is located by the vector

$$\bar{r} \doteq \frac{\sum_{i=1}^{n} (\rho_i \Delta_i V) r_i}{\sum_{i=1}^{n} \rho_i \Delta_i V}.$$

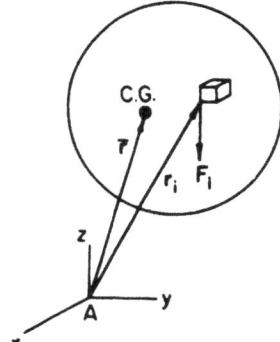

Fig. 98

As is indicated by the dot above the equal sign, this expression represents an approximation due to the fact that the weight of each element was assumed concentrated at a point. To obtain an exact expression it is necessary to increase the number of subdivisions indefinitely in such a manner that the volume of each tends to zero. The summation is then replaced by an integration over the volume, V, of the body, with the result

$$\bar{r} = \frac{\int_V \rho r \, dV}{\int_V \rho \, dV}.$$

The point defined by the position vector $\bar{r}$ is called the *center of gravity* of the body. It has the property of being a point on the line of action of the resultant gravitational force regardless of the orientation of the body. That is, a body subjected to the action of its own weight will be in static equilibrium if simply supported at its center of gravity.

If we refer the body to a set of rectangular axes with origin at A (Fig. 98), then the coordinates of its center of gravity (i.e., the components of the position vector $\bar{r}$) are

$$\bar{x} = \frac{\int_V \rho x dV}{\int_V \rho dV}, \quad \bar{y} = \frac{\int_V \rho y dV}{\int_V \rho dV}, \quad \bar{z} = \frac{\int_V \rho z dV}{\int_V \rho dV}.$$

It should be noted that, in general, the density is a (specified) function of the coordinates. For the case of a body of uniform density (i.e., a homogeneous body), the expression for $\bar{r}$ reduces to

$$\bar{r} = \frac{1}{V} \int_V r dV.$$

The center of gravity in such a case coincides with the *geometric center*, or *centroid*, of the body.

To illustrate the evaluation of the coordinates of the center of gravity it will suffice to treat examples involving only two dimensions. The volume in each integral is replaced by the area of the figure.

Example 1: Determine the coordinates of the centroid of the rectangular and triangular areas shown in Fig. 99.

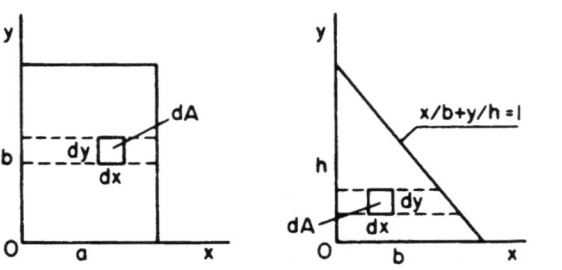

Fig. 99

(a) $\bar{x} = \dfrac{1}{A} \displaystyle\int_0^b \int_0^a x dx dy = \dfrac{a}{2}; \quad \bar{y} = \dfrac{1}{A} \displaystyle\int_0^b \int_0^a y dx dy = \dfrac{b}{2}.$

(b) $\bar{x} = \dfrac{1}{A} \displaystyle\int_0^h \int_0^{b(1-y/h)} x dx dy = \dfrac{b}{3}; \quad \bar{y} = \dfrac{1}{A} \displaystyle\int_0^h \int_0^{b(1-y/h)} y dx dy = \dfrac{h}{3}.$

CENTER OF GRAVITY

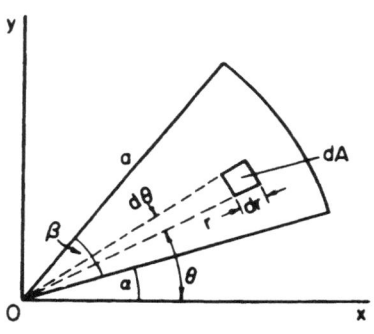

Fig. 100

Example 2: Compute the centroidal coordinates of the sector of a circle illustrated in Fig. 100.

$$\bar{x} = \frac{\iint x\,dA}{\iint dA} =$$

$$\frac{\int_\alpha^{\alpha+\beta}\int_0^a r^2 \cos\theta\,dr\,d\theta}{\int_\alpha^{\alpha+\beta}\int_0^a r\,dr\,d\theta}$$

$$= \frac{2a}{3\beta}[\sin(\alpha+\beta)-\sin\alpha];$$

$$\bar{y} = \frac{\iint y\,dA}{\iint dA} = \frac{\int_\alpha^{\alpha+\beta}\int_0^a r^2 \sin\theta\,dr\,d\theta}{\int_\alpha^{\alpha+\beta}\int_0^a r\,dr\,d\theta} = \frac{2a}{3\beta}[\cos\alpha - \cos(\alpha+\beta)].$$

In particular, for a sector bounded by the positive coordinate axes, $\alpha = 0$ and $\beta = \pi/2$, so that $\bar{x} = \bar{y} = 4a/3\pi$. For a semicircular sector bounded by the horizontal axis, $\alpha = 0$, $\beta = \pi$; hence, $\bar{x} = 0$, $\bar{y} = 4a/3\pi$.

In dealing with figures whose boundaries are not expressible by means of simple analytical relations but which are compounded of simple geometric shapes, it is often more convenient to subdivide the area into a finite number of elements, each having finite dimensions and known location of the centroid, and sum over these elements rather than integrate over the area of the figure. The procedure is illustrated below.

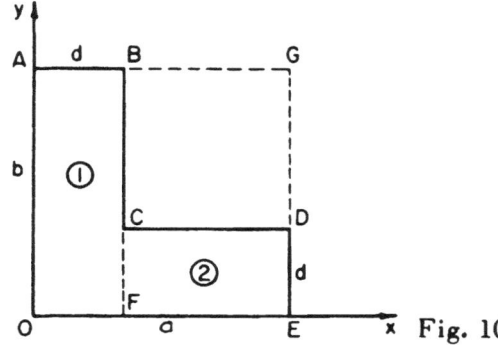

Fig. 101

Example 3: Find the coordinates of the centroid of the *L*-shaped area *OABCDEO* shown in Fig. 101.

The elements of finite dimensions may, in this case, be taken to be the rectangles *OABF* and *CDEF*, their area and centroidal coordinates being denoted by A_1, $\bar{x}_1$, $\bar{y}_1$ and A_2, $\bar{x}_2$, $\bar{y}_2$, respectively.

Accordingly,

$$\bar{x} = \frac{\bar{x}_1 A_1 + \bar{x}_2 A_2}{A_1 + A_2} = \frac{(d/2)(bd) + [d + (a-d)/2](a-d)d}{bd + (a-d)d} = \frac{bd + a^2 - d^2}{2(b + a - d)};$$

$$\bar{y} = \frac{\bar{y}_1 A_1 + \bar{y}_2 A_2}{A_1 + A_2} = \frac{(b/2)(bd) + (d/2)(a-d)d}{bd + (a-d)d} = \frac{b^2 + ad - d^2}{2(b + a - d)}.$$

Alternatively, we may consider the elements of the area in Fig. 101 to be the rectangles $OAGE$ and $BGDC$; the latter, however, must be treated as having negative area in order for the total area to be that of the desired figure. Proceeding in this manner, we get

$$\bar{x} = \frac{(a/2)(ab) + [d + (a-d)/2][-(a-d)(b-d)]}{ab + [-(a-d)(b-d)]} = \frac{bd + a^2 - d^2}{2(b + a - d)};$$

$$\bar{y} = \frac{(b/2)(ab) + [d + (b-d)/2][-(a-d)(b-d)]}{ab + [-(a-d)(b-d)]} = \frac{b^2 + ad - d^2}{2(b + a - d)}.$$

3. Application to Problems Involving Distributed Load

An example of a structural member carrying distributed load is furnished by the beam illustrated in Fig 102. Supported on smooth knife edges at A and B, it is subjected to a normal load of magnitude w per unit length. The support reactions are obtained from the equations of equilibrium:

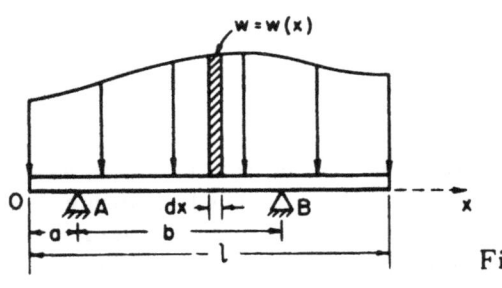

Fig. 102

$$\Sigma F_y = R_A + R_B - \int_0^l w(x)dx = 0,$$

$$\Sigma M_O = aR_A + (a + b)R_B - \int_0^l xw(x)dx = 0.$$

We note that the integral in the first equation represents the area under the curve $w = w(x)$, i.e., the total load, W, while the integral in the second equation is proportional to the x-coordinate of the centroid of this area. Consequently, these equations may be written

$$R_A + R_B - W = 0, \quad aR_A + (a+b)R_B - W\bar{x} = 0,$$

with the result

$$R_A = W\left(1 - \frac{\bar{x} - a}{b}\right), \quad R_B = W\left(\frac{\bar{x} - a}{b}\right).$$

A similar computation involving loads distributed over the surface of a plate gives rise to integrals which may be interpreted as being proportional to centroidal coordinates of the volume capped by the loading surface $w = w(x,y)$.

Example: Determine the reaction at the supports of the beam loaded as shown in Fig. 103.

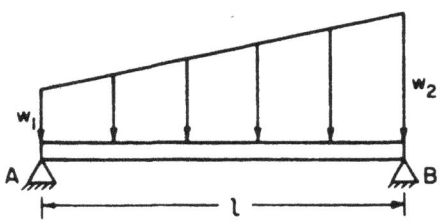

Fig. 103

The variation of the load is expressed by the equation

$$w(x) = w_1 + (w_2 - w_1)x/l;$$

hence, the total load is given by

$$W = \int_0^l w(x)dx$$
$$= (w_1 + w_2)l/2,$$

and the centroidal coordinate, by

$$\bar{x} = \frac{1}{W}\int_0^l xw(x)dx = \frac{l}{3}\left(\frac{w_1 + 2w_2}{w_1 + w_2}\right).$$

Accordingly, the reactions are of magnitude

$$R_A = W(1 - \bar{x}/l) = (2w_1 + w_2)l/6,$$
$$R_B = W\bar{x}/l = (w_1 + 2w_2)l/6.$$

PROBLEMS

Chapter I. Elements of Vector Algebra

1. Find the magnitude and direction of the vectors whose scalar components in a rectangular coordinate system are: (a) 0, 5, 12; (b) 3, 4, 5; (c) 3, 3, 3; (d) 1, −2, 3; (e) −2, −4, 1.

2. The magnitude of vector **A** is $A = 47$ and its direction angles are $\alpha = 129.5°$, $\beta = 71°$, $\gamma > 90°$. Vector **B** is of magnitude $B = 127$ and its line of action passes through points P_1 and P_2, in the sense from P_1 to P_2, whose coordinates with respect to an arbitrary fixed origin are (−1, 2, 3) and (0, 4, 1), respectively.
Compute the magnitude and direction of the vectors (a) **A** + **B**, (b) **A** − **B**, (c) 2**A** − 3**B**.

3. Two vectors are specified in terms of unit coordinate vectors by means of the expressions **P** = −**i** + 2**j** − 3**k**, **Q** = −4**i** + 3**j** + **k**. Find (a) the unit vectors along **P**, **Q**, **P** + **Q**; (b) the scalar product **P** · **Q**; (c) the scalar component of **Q** in the direction − **P**; (d) the angle between **P** and **Q**.

4. Show that of the vectors

$$\mathbf{a} = (T \cos \theta)\mathbf{i} + (T \sin \theta)\mathbf{j}$$
$$\mathbf{b} = (S \sin \theta)\mathbf{i} - (S \cos \theta)\mathbf{j}$$
$$\mathbf{c} = (P \cos \theta)\mathbf{i} + (P \sin \theta)\mathbf{j}$$

(a) **a** and **b** are perpendicular; (b) **a** and **c** are parallel.

5. A vector is given in terms of its cartesian components by the expression **P** = **i** − 2**j**. The line of action of **P** passes through point A whose coordinates with respect to an origin located at O are (0, −1, 1).
(a) Compute the moment, **M**, of **P** about O. (b) Find the moment of **P** about the y-axis.

PROBLEMS

6. The line of action of vector **R**, whose magnitude is 10, passes through points A and B whose rectangular coordinates with respect to origin O are $(1, -2, -1)$ and $(1, -1, 1)$, respectively.

Determine the moment of **R** (a) about point $C(-1, 1, 0)$, and (b) about line L which passes through C and B in the direction CB.

7. Find the moment of a vector **A**, whose line of action is directed along a body diagonal of a cube of edge-length b, about any face diagonal of the cube which does not intersect the line of action of the vector.

8. Three unit vectors converge on a corner of a cube along three edges. Find the moment, about the diagonally opposite corner of the cube, of each of these vectors, and of their sum.

9. A vector **R** whose components are (a, b, c) acts at a point D (x, y, z). What is its moment about a line L through the origin with direction cosines (l, m, n)?

10. The lines of action of each of three vectors P_1, P_2, P_3 coincide with the diagonals of three adjacent faces of a cube of edge-length unity, all passing through the same corner O and directed away from it.

Referring the system to coordinate axes with origin at O such that they are coincident with the edges of the cube, find the moment of the sum $(P_1 + P_2 + P_3)$ about (a) the corner A $(1, 0, 0)$; (b) the body diagonal through A.

11. A line L having direction cosines (l, m, n) passes through point (a, b, c). Find the moment about L of a unit vector pointing along the x-axis.

Chapter II. The Problem of Equilibrium

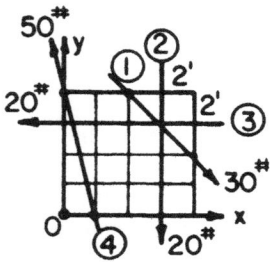

12. For the system shown, determine the magnitude and direction of (a) the resultant force, (b) the resultant moment about O.

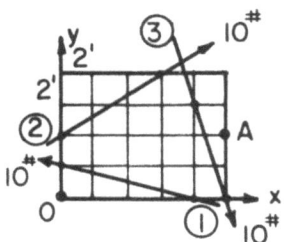

13. Compute the magnitude and direction of (a) the resultant force, (b) the resultant moment about O, (c) the resultant moment about A.

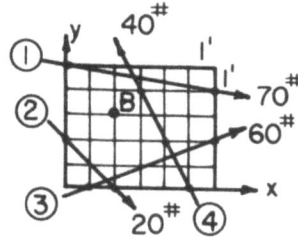

14. Find (a) the resultant force, (b) the resultant moment about B.

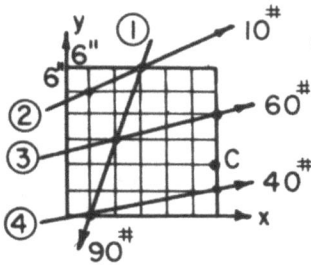

15. Determine the resultant force and the resultant moment about C.

PROBLEMS

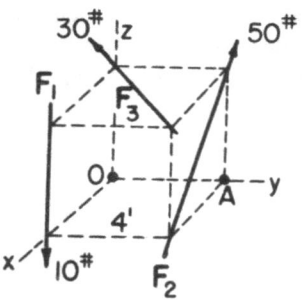

16. Three forces act on a cube whose edge is 4 ft. in length, as shown. Find (a) the resultant force, (b) the resultant moment about O, (c) the resultant moment about A.

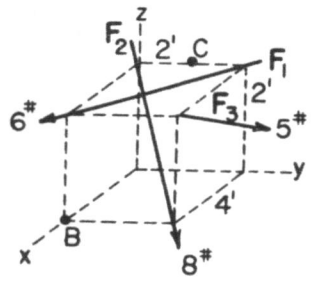

17. The cube of problem 16 is acted upon by three other forces, shown in the diagram. Compute the resultant moment about (a) point B, (b) point C.

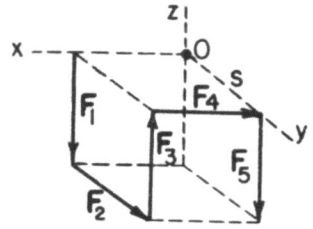

18. Five forces act along the edges of a cube of side s ft. The magnitude of each of the odd numbered forces is p lbs. and of the even numbered ones it is q lbs. What is the resultant moment about (a) the origin O? (b) the center of the front face?

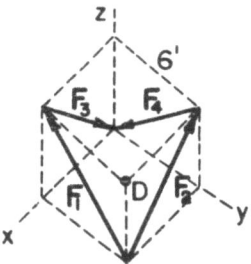

19. The cube shown in the diagram has four forces acting along face diagonals. The magnitude of each force, in lbs., equals ten times its subscript. Compute (a) the resultant force, (b) the total moment about D.

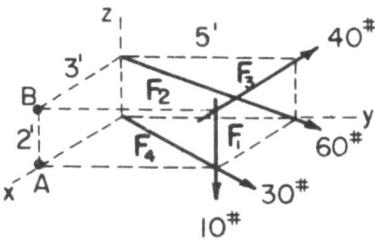

20. For the system of four forces shown in the figure, find (a) the resultant force, (b) the resultant moment about A, (c) the resultant moment about B.

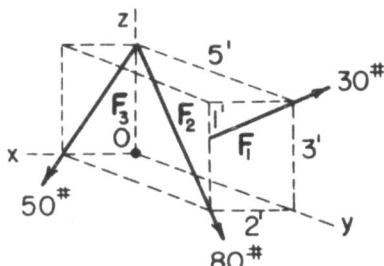

21. Determine (a) the resultant force and (b) the resultant moment about O for the system shown.

PROBLEMS

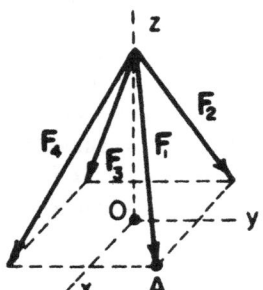

22. Four forces, each of magnitude 100 lbs., act at the vertex of a regular pyramid along each of four edges, as shown. The base of the pyramid is a square 2 ft. × 2 ft., its altitude is 4 ft. Find the moment (a) of each force about the center of the base, point O; (b) of the resultant force about corner A.

23. Suppose that, in problem 22, each force is of magnitude P lbs. and the pyramid has a rectangular base of sides a and b, parallel to x and y, respectively, and altitude h (all in ft.). Determine the moment of each force about (a) the origin O, (b) the point of intersection of the positive x-axis with the side of the base.

Chapter III. Equilibrium of Simple Planar Systems

24. Two blocks, A and B, each weighing 100 lbs., are suspended from a fixed point C by two strings, AC and BC, each 14 ft. long. The blocks are separated by a stick 4 ft. in length. Neglecting the weight of the stick, find the tension in the strings and the thrust in the stick.

Ans. $T_{AC} = 101.0$ lbs.
$C_{AB} = 14.4$ lbs.

25. A man walking a tight-wire stops at the one-third point between the supports. It is noted that the deflection at this point is 5 ft. The man weighs 150 lbs. and the distance between supports is 60 ft. Assuming the weight of the wire to be negligible, compute the tension in the shorter portion of the wire.

Ans. $T = 412$ lbs.

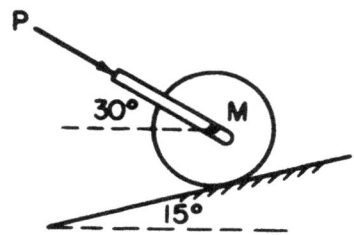

26. Find the force P which must be applied to the 200 lb. lawn mower, M, in order to roll it up the incline at constant velocity.

Ans. $P = 73.3$ lbs.

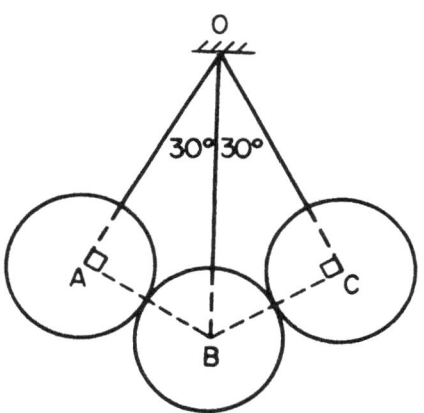

27. Three smooth spheres, A, B, and C, of equal diameters, are suspended by cords from a fixed point O, as shown in the figure. A and C weigh 50 lbs. each, B weighs 100 lbs., and the centers of the spheres lie in a vertical plane. Find the tension in the cord OB.

Ans. $T_{OB} = 125$ lbs.

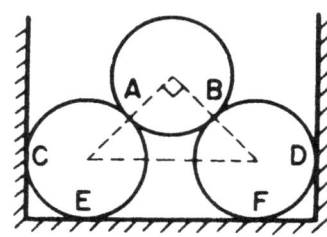

28. Three smooth cylinders, having equal radii and each weighing 1000 lbs., are stacked in a container, as shown in the diagram. Find the pressure at all contacts.

Ans. $A = B = 707$ lbs.
$C = D = 500$ lbs.
$E = F = 1500$ lbs.

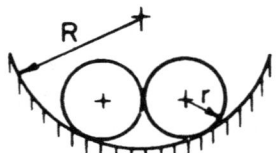

29. Two like, smooth cylinders, each of radius r and weight W, are placed in a fixed, smooth cylindrical container of radius R, as shown. Compute the contact force between the cylinders.

Ans. $C = Wr\,(R^2 - 2Rr)^{-1/2}$

30. A third like cylinder is added to the system of problem 29, such that it rests symmetrically on top of the first two. Find the largest radius R of the container for which the cylinders will remain in contact.

Ans. $R = 6.29r$

31. A particle in contact with a smooth inclined plane may be held in equilibrium either by a horizontal force of 100 lbs. or by a force of 75 lbs. inclined to the plane by an angle of $45°$. Compute the weight of the particle and the angle of the plane with the horizontal.

Ans. $\alpha = 58°$
$W = 62.5$ lbs.

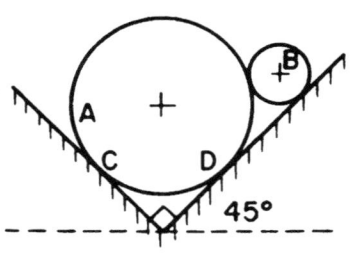

32. Two smooth cylinders are placed in a smooth $90°$-trough, as shown. Cylinder A has a diameter of 6 in. and weighs 20 lbs., cylinder B has a diameter of 2 in. and weighs 10 lbs. Find the pressure on A at C and D.

Ans. $N_C = 21.2$ lbs.
$N_D = 10.0$ lbs.

33. Two spheres, S_1 and S_2, of weight W_1 and W_2, respectively, are connected by a rod of length l and of negligible weight and placed on a pair of smooth inclined planes, as shown in the diagram.

Compute the angle θ between the rod and the horizontal when the system is in equilibrium for the cases (a) $\alpha = \beta$, $W_1 > W_2$; (b) $W_1 = W_2$, $\alpha > \beta$.

Ans. (a) $\tan \theta = \dfrac{W_1 - W_2}{W_1 + W_2} \cot \alpha$

(b) $\tan \theta = \frac{1}{2}(\cot \beta - \cot \alpha)$

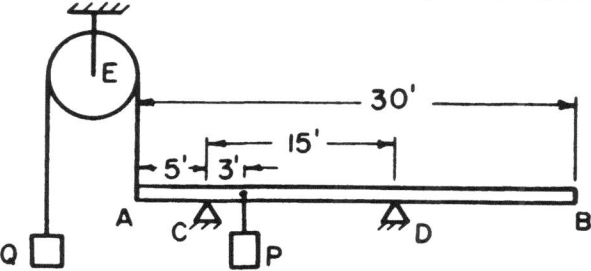

34. A uniform beam AB, 30 ft. long and weighing 15 lbs. per ft., rests on supports C and D, 15 ft. apart. A weight Q of 500 lbs. is suspended from a string which passes over pulley E, of negligible weight, and which is attached to the beam at point A, 5 ft. to the left of support C. A weight $P = 1500$ lbs. is suspended from the beam at a point 3 ft. to the right of C. Compute the reaction of the supports.

Ans. $C = 683$ lbs.
$D = 767$ lbs.

35. A beam weighing W lbs. and l ft. long is suspended in a horizontal position by two cables attached at its ends. How far from one end must a load of $2W$ lbs. be placed on the beam in order that one of the cables be subjected to twice the tension in the other?

Ans. $x = l/4$

36. When placed in one pan of a balance, a body weighs 10 lbs., and when placed in the other pan, it weighs 10.1 lbs. Determine the true weight of the body, and the ratio of the lengths of the arms of the balance.

Ans. $W = \sqrt{101}$ lbs., $a/b = 10/\sqrt{101}$

37. A uniform bar is placed on a table with one end projecting 1 ft. over the edge. If a 10 lb. weight is hung on the projecting end of the bar, the bar is then on the verge of tilting. If an additional weight of 30 lbs. is placed on the supported end of the bar, the bar may project 3 ft. over the edge. Determine the length and weight of the bar.

Ans. $W = 8.85$ lbs.
$l = 4.26$ ft.

PROBLEMS

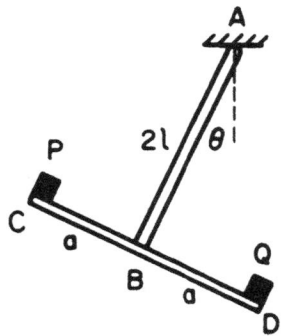

38. Two uniform rods, AB and CD, of negligible weight and of length $2l$ and $2a$, respectively, are welded at B to form a right angle, and the system is freely suspended by means of a pivot at A. Two blocks, of weight P and Q, are then placed at the ends C and D, as shown. Assuming $Q > P$, compute the angle θ at equilibrium.

$$\text{Ans. } \tan \theta = \frac{a(Q - P)}{2l(Q + P)}$$

39. Solve problem 38 if the weight of rods AB and CD is W_1 and W_2, respectively.

$$\text{Ans. } \tan \theta = \frac{(a/l)(Q - P)}{W_1 + 2W_2 + 2(Q + P)}$$

40. A solid hemisphere of radius r and weight W rests with its convex surface on a smooth horizontal table. A particle of weight P is placed on the extremity of a diameter of the plane face of the hemisphere. Determine the angle θ between this face and the table when the system is in equilibrium.

Note: the center of gravity of a hemisphere is located at a height $3r/8$ above the base.

$$\text{Ans. } \theta = \text{arc tan } (8P/3W)$$

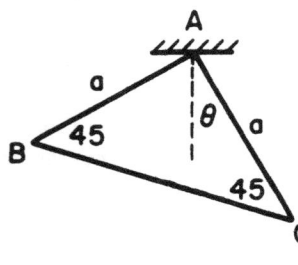

41. Three uniform rods are welded into the form of an isosceles right triangle ABC which is then freely suspended from a hinge at A. The weight of AB is W_1; of AC, W_2; of BC, negligible. The dimensions are as shown in the figure. Compute the angle θ which the rod AC makes with the vertical in a position of equilibrium.

$$\text{Ans. } \theta = \text{arc tan } (W_1/W_2)$$

42. Determine the angle θ in problem 41 if the weight of rod BC is W_3.

$$\text{Ans. } \tan \theta = \frac{W_1 + W_3}{W_2 + W_3}$$

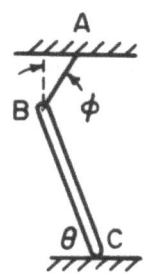

43. The lower end C of a uniform bar BC, 12 ft. long and weighing 200 lbs., rests on a smooth floor and the upper end B is fastened to the ceiling by means of string AB, 6 ft. long, as shown in the figure. If the height of the ceiling is 10 ft., compute the tension in the string, the angle ϕ which the string makes with the vertical, the reaction of the floor on the bar, and the angle of inclination, θ, of the bar to the floor.

Ans. $\theta = 19°27'$

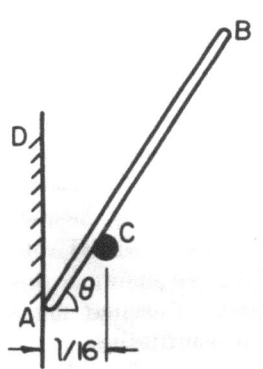

44. A uniform rod AB, of length l and weight W, rests against a smooth peg C and is in contact with a smooth vertical wall AD, as shown. If the horizontal distance between the peg and the wall is $l/16$, determine the angle θ which the bar makes with the horizontal at equilibrium.

Ans. $\theta = 60°$

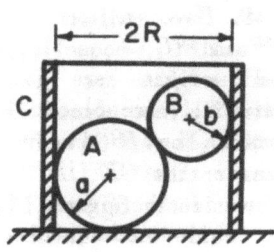

45. A hollow circular tube C, open on both ends, is placed on a table and two spheres, A and B, of weight P and Q respectively, are inserted, as shown in the figure. Assuming all contacts to be smooth, compute the least weight of the tube necessary to prevent tipping.

Ans. $W_{min.} = Q \dfrac{2R - a - b}{R}$

PROBLEMS

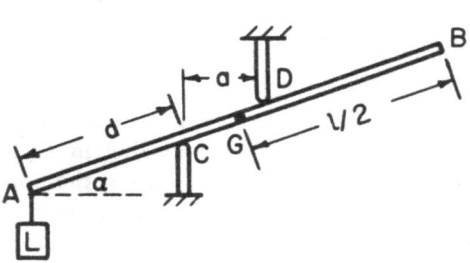

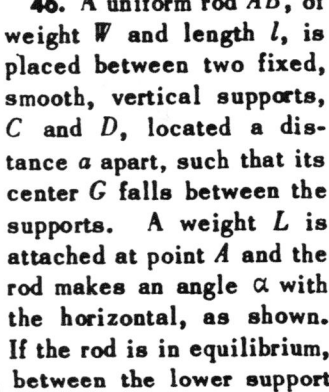

46. A uniform rod AB, of weight W and length l, is placed between two fixed, smooth, vertical supports, C and D, located a distance a apart, such that its center G falls between the supports. A weight L is attached at point A and the rod makes an angle α with the horizontal, as shown. If the rod is in equilibrium, find the distance d, measured along the rod, between the lower support and the lower end of the rod.

Ans. $d = \dfrac{l}{2}\left(\dfrac{W}{W+L} + \dfrac{a}{l \cos^3 \theta}\right)$

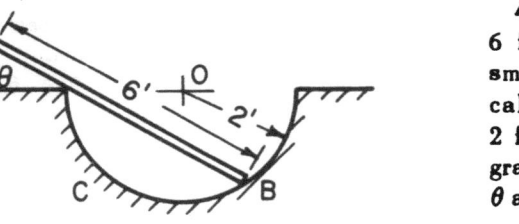

47. A uniform bar AB, 6 ft. long, is placed in a smooth, fixed, hemispherical bowl C whose radius is 2 ft., as shown in the diagram. Compute the angle θ at equilibrium.

Ans. $\theta = 23°13'$

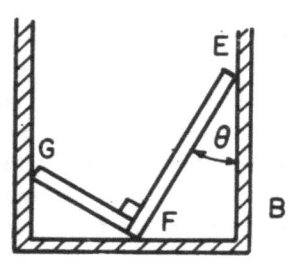

48. Two uniform rods, EF and FG, whose lengths and weights are in the ratio 2:1, are placed in a smooth box B in such a manner that $GF \perp EF$, as shown in the figure. Find the angle θ at equilibrium.

Ans. $\theta = 35°16'$

49. Suppose that, in problem 48, EFG represents a single bar bent to form a right angle, and that, in the position of equilibrium, $\theta = 30°$. Determine the contact forces at E, F, G.

Ans. $E = 0.46 W$

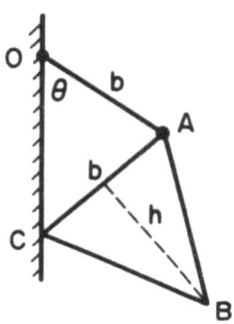

50. A uniform plate ABC, in the shape of an isosceles triangle of base b and altitude h, is suspended by means of string OA, of length b, from a smooth vertical wall, with corner C resting in contact with the wall. Determine the angle θ which the string makes with the wall at equilibrium.

Ans. $\tan \theta = 2h/9b$

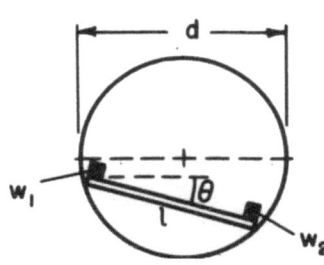

51. Weights W_1 and W_2 are fastened at the extremities of a slender, uniform rod of length l and the assembly is then placed inside a fixed, smooth, spherical shell of diameter d, as shown. Find the angle θ which the rod makes with the horizontal in a position of equilibrium (a) assuming the weight of the rod to be negligible, (b) taking into account the weight, W, of the rod.

Ans. (b) $\tan \theta = \dfrac{W_2 - W_1}{W + W_1 + W_2} \cdot \dfrac{l}{(d^2 - l^2)^{1/2}}$

52. Compute the magnitude of the contact reactions in problem 51 if $W = W_1 = 10$ lbs., $W_2 = 20$ lbs., $l = 12$ in., $d = 13$ in.

Ans. $N_1 = 33.45$ lbs., $N_2 = 55.75$ lbs.

Chapter IV. Equivalence of Force Systems

Compute the magnitude, direction, and the points of intersection of its line of action with the coordinate axes, of the single force which is statically equivalent to the system of forces shown in the diagram corresponding to

PROBLEMS

53. Problem 12.

54. Problem 13.

55. Problem 14.

56. Problem 15.

For the system of forces pertaining to each of the following problems, find the resultant moment about the origin of coordinates and then use this quantity to determine the resultant moment about

57. Point A, problem 16.

58. Point B, problem 17.

59. Point D, problem 19.

60. Point (1, 1, 1), problem 20.

61. Point (−1, 2, −1), problem 21.

62. Point (2, 2, −1), problem 22.

Chapter V. Simple Structures

For the trusses illustrated below, determine, by employing the method of joints, the force in each member.

63.

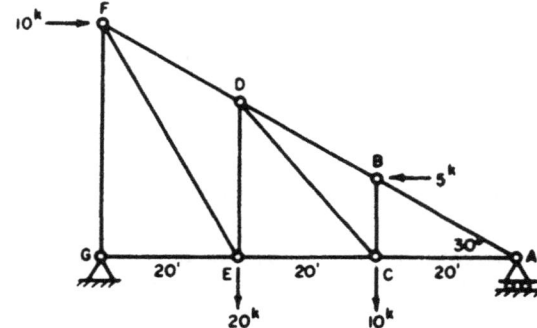

PROBLEMS

64.
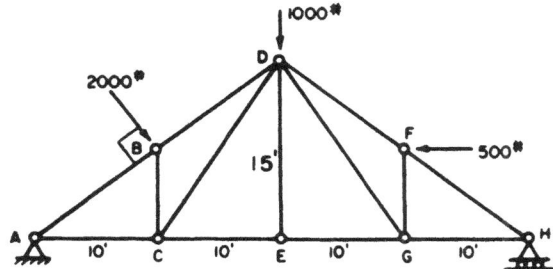

65
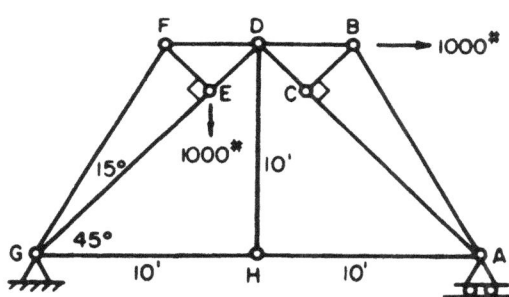

66.
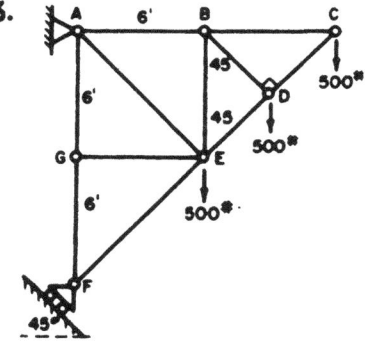

67.
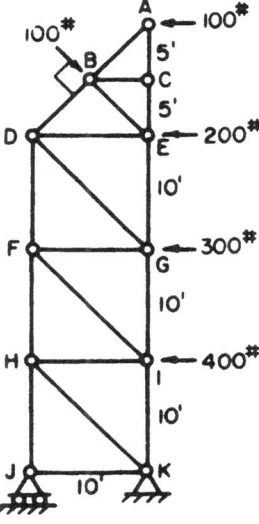

68.

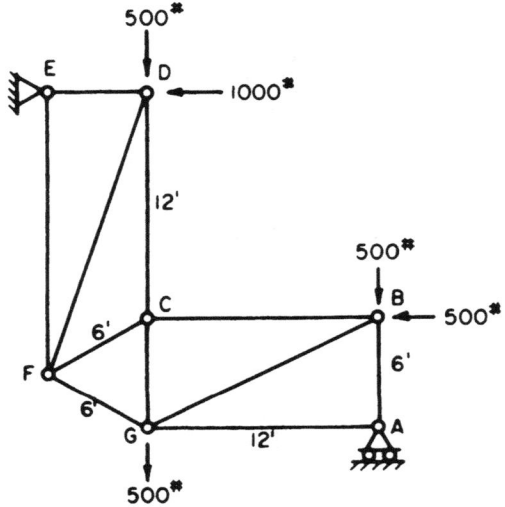

69.

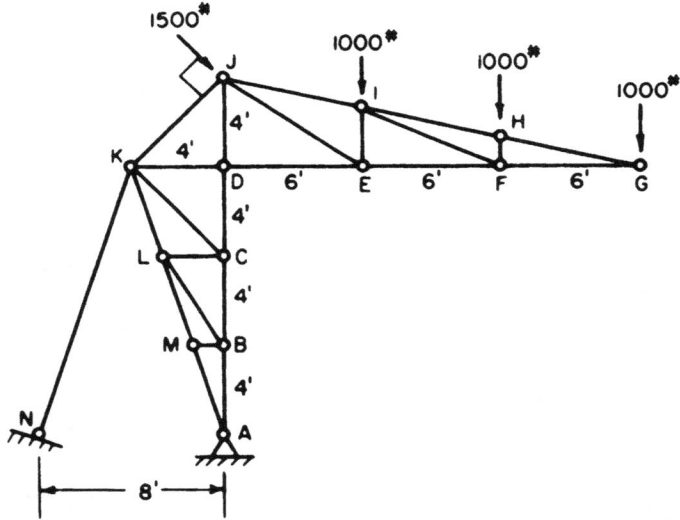

Use the method of sections to compute the force in members

70. *DC* and *DE* of problem 63.

71. *DC* and *DG* of problem 64.

72. *DC* and *DE* of problem 65.

73. *BE* and *DE* of problem 66.

74. *BE*, *FH*, and *FI* of problem 67.

75. *CF* and *FG* of problem 68.

76. *JE* and *KC* of problem 69.

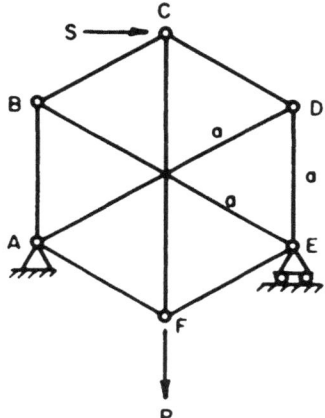

77. Find the force in each member of the complex truss shown in the figure due to (a) the vertical force P applied at F, (b) the horizontal force S applied at C.

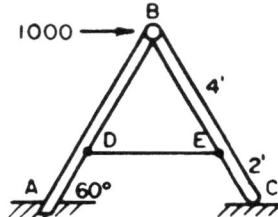

78. Two uniform bars, AB and BC, are hinged at B and supported at A and C to form a symmetric frame. The structure is held in equilibrium under the applied load by means of the horizontal cable DE, as shown. Assuming the contact at C to be smooth, and the weight of the bars to be 100 lbs./ft., compute the magnitude and direction of the reactions on member AB at A and B.

79. Suppose that, in problem 78, the cable DE is replaced by a uniform, rigid member weighing 200 lbs. Determine the reactions on bar AB at B, D and A.

PROBLEMS

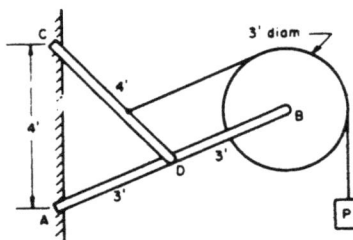

80. Determine the magnitude and direction of the reactions on CD at C and D if $P = 500$ lbs. and AB, CD each weigh 25 lbs./ft.

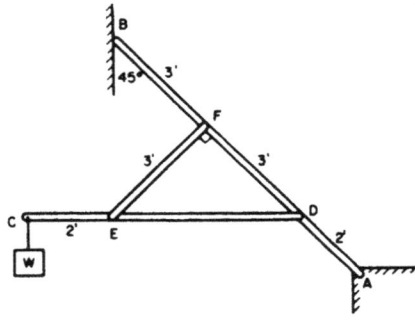

81. Compute the magnitude and direction of all reactions on AB if $W = 1000$ lbs. and the weight of the members is neglected.

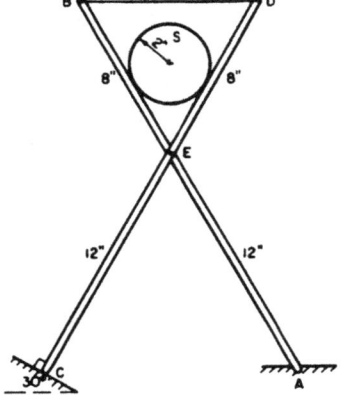

82. Compute the reactions at A and E if sphere S weighs 10 lbs. and the weight of the members is assumed negligible. Note that cable BD is horizontal.

PROBLEMS

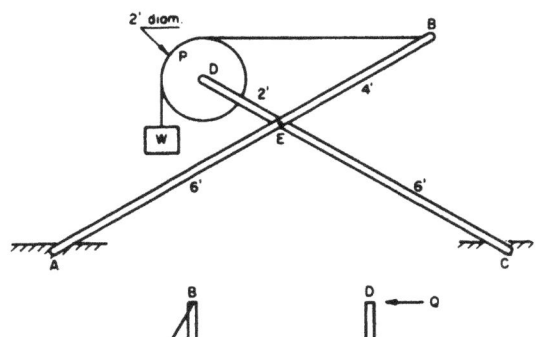

83. Find the reactions at A and E (on AB) if the load $W = 600$ lbs., the pulley P weighs 100 lbs., and each member weighs 25 lbs./ft.

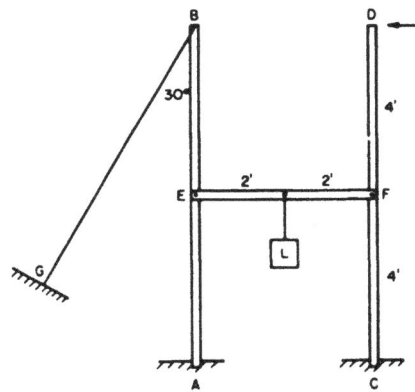

84. In the frame shown in the figure, find the reactions at E and F, on bar EF, for $L = 200$ lbs. and $Q = 500$ lbs. Neglect the weight of the members.

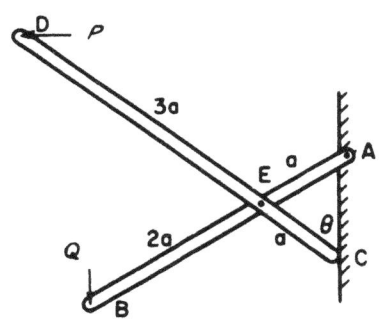

85. Determine the magnitude and direction of the reaction on member AB at E and the value of angle θ at equilibrium (a) assuming the weight of bars AB and CD to be negligible, (b) taking into account that AB and CD are uniform and of weight w per length a.

Ans. (a) $\theta = \arctan(4P/3Q)$
(b) $\theta = \arctan \dfrac{4}{3}\left(\dfrac{2P}{2Q - 3wa}\right)$

Chapter VI. Sliding Friction

86. Bracket B, weighing 20 lbs., is hinged by means of a smooth pin at A and is in contact ($\mu = 0.2$) at C with slab S weighing 35 lbs. The slab, in turn, rests on a rough floor ($\mu = 0.3$). Compute the magnitude of the least horizontal force P which will cause impending motion of slab S if a load $Q = 100$ lbs. is applied to the bracket as shown in the figure.

Ans. $P = 53$ lbs.

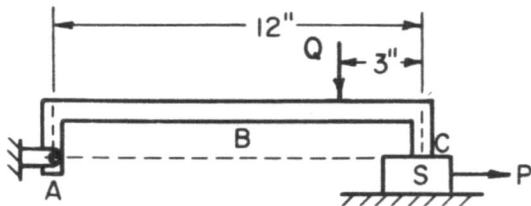

87. A force $Q = 300$ lbs. is applied to the pulley shown in the diagram. Compute the least force P, applied at the end of the brake lever, which will prevent rotation of the pulley if the coefficient of friction at the brake contact is $\mu = 0.5$

Ans. $P = 56.25$ lbs.

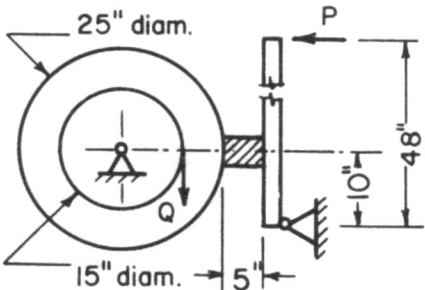

88. A block of weight W rests on a rough horizontal floor, the coefficient of static friction being μ. It is desired to move the block by pulling on a cable attached to it. Find the least tension in the cable which will cause impending motion, and the angle which the cable makes with the floor.

Ans. $P_{min.} = \mu W (1 + \mu^2)^{-1/2}$

89. A particle of weight W is placed on a fixed spherical surface of radius a. If the coefficient of static friction at the contact is μ, compute the greatest distance from the top of the vertical diameter at which the particle will remain in equilibrium.

Ans. $a \tan^{-1} \mu$

90. In the system of problem 43 the contact between the bar and the floor is assumed to be rough. If the angle θ which the bar makes with the floor at impending motion is $30°$, compute the coefficient of static friction of the contact.

Ans. $\mu = 0.488$

PROBLEMS

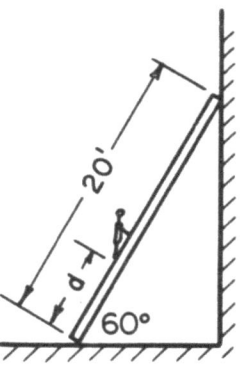

91. Compute the greatest height d to which a man weighing 150 lbs. can ascend along the 75 lb. ladder shown in the diagram if the coefficient of friction at the contacts with the floor and the wall is $\mu = 0.3$.

Ans. $d = 11.77$ ft.

92. In problem 91, what is the least angle of inclination of the ladder to the floor which will enable the man to ascend to the top?

Ans. $\theta = 70°12'$

93. If, in problem 44, the contact at A is a rough contact (the contact at C being smooth), and the angle θ at impending motion is $55°$, find the coefficient of friction.

Ans. $\mu = 0.237$

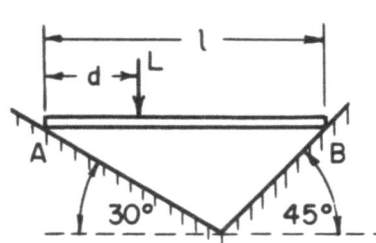

94. A horizontal plank AB, of negligible weight, is in equilibrium with its ends resting on fixed inclined planes, as shown. If the coefficient of static friction at each contact is 0.25, compute the closest distance, d, as a fraction of the length, l, between the applied load L and end A for which no slipping occurs. How close can the load be placed to the end B before slipping takes place?

Ans. $d_A/l = 0.146$
$d_B/l = 0.383$

95. Block B, of weight W, rests on a rough, inclined plane. Link AC, of length l and negligible weight, is rigidly attached to the block and a

118 **PROBLEMS**

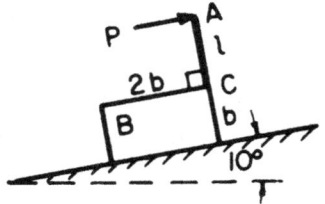

force of magnitude P applied at A parallel to the plane, as shown. If the coefficient of static friction at the contact is 0.2, compute the greatest length l for which the block will tend to slide up the plane rather than tip over, and the force necessary to accomplish this.

Ans. $\dfrac{l}{b} = 1.9$, $\dfrac{P}{W} = .371$

96. Determine the greatest length l in problem 95 at which sliding will occur without tipping if the applied force P is horizontal; what is the magnitude of the corresponding force?

Ans. $\dfrac{l}{b} = 1.8$, $\dfrac{P}{W} = .390$

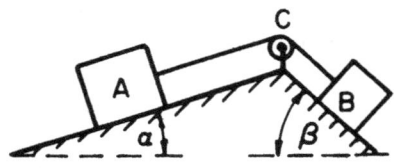

97. Blocks A and B, of weight W_A and W_B, respectively, rest on fixed inclined planes and are connected by means of a string which passes over a smooth pulley, C, as shown. If the coefficient of static friction at each contact is μ, compute (a) the greatest, (b) the least, value of the ratio of the weights, consistent with static equilibrium. (c) Suppose that, for $\alpha = 30°$, $\beta = 45°$, and $W_B = 1.5 W_A$, motion impends such that B tends to move down the incline; determine the coefficient of friction.

Ans. (a) $\dfrac{W_A}{W_B} = \dfrac{\sin \beta + \mu \cos \beta}{\sin \alpha - \mu \cos \alpha}$ (providing $\tan \alpha > \mu$)

(c) $\mu = 0.206$

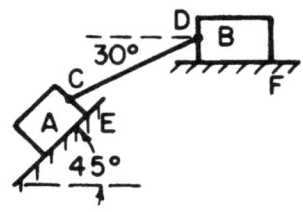

98. Block B, weighing 100 lbs., is in contact with horizontal plane F ($\mu_F = 0.25$), and is connected by means of cable CD to block A, weighing 50 lbs., which, in turn, is in contact with inclined plane E ($\mu_E = 0.2$), as shown in the figure.

(a) Compute the tension in CD and determine whether the system will move. (b) Find the least coefficient of friction μ_F for which the system will remain at rest, assuming all other quantities unchanged.

Ans. (b.)$\mu_F = 0.211$

99. In problem 98, what is the least magnitude of the horizontal force which, when applied to block B, will cause sliding of A up plane E?

Ans. $P = 70.99$ lbs.

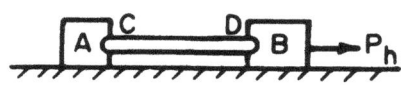

100. Crates A and B, of weight W_A and W_B, respectively, rest on a horizontal floor and are connected by a horizontal rigid link CD of weight W. The coefficient of static friction at the contacts has the values μ_A and μ_B, respectively.

Determine the least horizontal force P_h required to cause sliding of the system.

Ans. $P_h = \mu_A W_A + \mu_B W_B + \frac{1}{2}(\mu_A + \mu_B) W$

101. In problem 100, what is the least force, P, necessary to cause sliding, and at what angle θ is it inclined to the floor?

Ans. $P = P_h (1 + \mu_B)^{-1/2}$,
$\theta = \arctan \mu_B$

102. A uniform plank, 16 ft. long and weighing 60 lbs., rests in a horizontal position on a fixed circular cylinder whose axis is horizontal and whose diameter is 4 ft. The plank is to be used as a see-saw. If the coefficient of static friction at the contact is 0.4, and a child weighing 40 lbs. is seated on one end, determine (a) the greatest, (b) the least, weight of the child on the other end consistent with no slipping of the plank.

Ans. (a) 54.8 lbs.
(b) 27.8 lbs.

Chapter VII. Work and Energy Methods

103. A chain l ft. long and weighing q lbs./ft. is coiled on the floor. Compute the work required to raise one end slowly to a height of h ft, if (a) $h < l$, (b) $h > l$.

104. A spring whose stiffness is k (constant) has one end attached to a fixed vertical bar at C and the other end connected to the mid point, D,

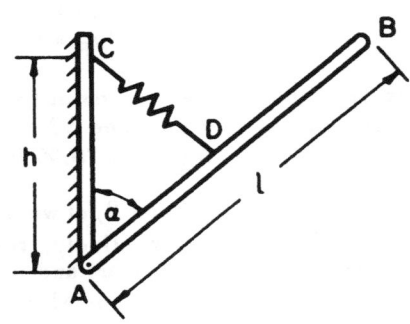

of a uniform bar AB. With AB pivoted about A, the undeformed position of the spring is shown in the diagram.

(a) Compute the work done in stretching the spring to the position in which point B is vertically below point A.
(b) Find the work done by the weight, W, of the bar AB in the course of this displacement. (c) Determine the answer to part (a) if the stiffness of the spring is numerically equal to twice its elongation.

105. The force exerted on a particle in the earth's gravitational field is always directed toward the center of the earth and its magnitude is $F = Wa^2/x^2$, where a is the radius of the earth, x is the distance of the particle from the center of the earth, and W is the weight of the particle on the surface of the earth. Compute the work which must be done against this force to enable the particle, initially on the earth's surface, to escape the earth's gravitational pull.

Ans. Work $= Wa$

106. A particle, P, is constrained to displace along a circular loop of wire, as shown, under the action of several forces.

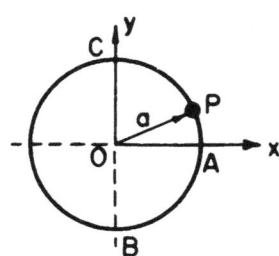

(a) The magnitude of one of these forces is numerically equal to twice the height of the particle above the base, B, of the vertical diameter; this force is always horizontal and directed toward the vertical diameter.

(b) A second force is such that its horizontal component is numerically equal to the sum, its vertical component to the product, of the distances to the coordinate axes.

(c) A third force is always directed toward the center, O, its magnitude being numerically equal to the tangent of the angle the radius to the particle makes with the horizontal.

Compute the work done by each force in the course of displacement from the extreme right, A, to the top, C, of the wire.

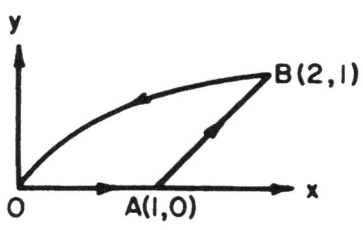

107. A particle displaces to point B along the straight-line paths OA and AB, and returns to O along the parabola BO, as shown in the diagram.

Compute the work done in the course of this displacement by
(a) $F_x = x^2 + y^2$, $F_y = x^2 - y^2$;
(b) $F_x = (x + y)^2$, $F_y = (x - y)^2$;
(c) $F_x = 2xy$, $F_y = x^2 - 2y^2$.

Ans. (c) Work = 0

108. A particle displaces along the circular helix $x = 2 \cos t$, $y = 2 \sin t$, $z = 3t$ from a position on plane $z = 3$ to one on plane $z = 10$. Compute the work done by a force whose magnitude is numerically equal to the height above the plane $z = 0$ and whose line of action at every point is normal to the axis of the helix and directed toward it.

Ans. Work = 0

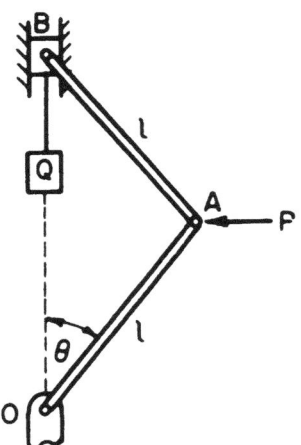

109. Two uniform bars OA and AB, each of length l and weight W, are pinned at A and connected, respectively, by means of hinges, to a fixed bar at O and a slider at B, as shown. The slider is subjected to a freely suspended load Q and is unrestrained in the vertical direction. Employ the principle of virtual work to compute the magnitude, P, of the horizontal force at A which will insure equilibrium with OA inclined at an angle θ to the vertical.

Ans. $P = 2(Q + W) \tan \theta$

PROBLEMS

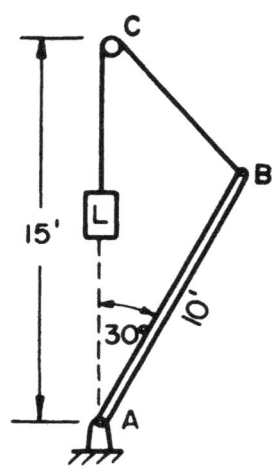

110. A uniform rod, AB, 10 ft. long and weighing 100 lbs., is pinned at A and restrained by a cable tied at B and passing over a smooth pulley, C, as shown. Apply the principle of virtual work to compute the load L required to maintain equilibrium at an angle of 30° with the vertical.

Ans. $L = 26.9$ lbs.

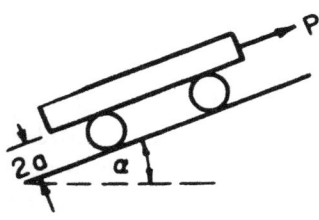

111. A beam of weight W rests on rollers of weight W_1 and radius a and the system is placed on an inclined plane, as shown. Compute, by means of the principle of virtual work, the magnitude, P, of the force, applied parallel to the incline, which will maintain equilibrium.

Ans. $P = (W + W_1) \sin \alpha$

112. The square frame shown in the diagram is fastened by smooth pins and supported by a 30° inclined plane at A and D. Apply the principle of virtual work to find the magnitude of the reaction in member BD

(a) if the weight of the members is neglected, (b) if each of the bars of length a weighs 100 lbs.

Ans. (a) R = 1414 lbs. (compression)

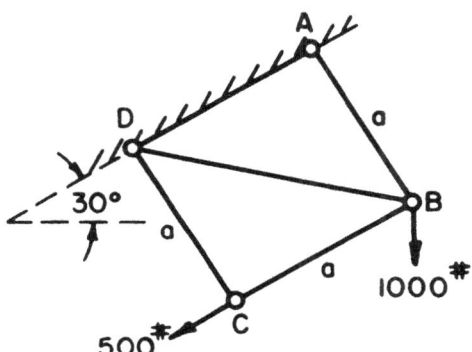

Solve the following by means of the principle of virtual work:

113. Problem 34.

114. Problem 38.

115. Problem 41.

116. Problem 85.

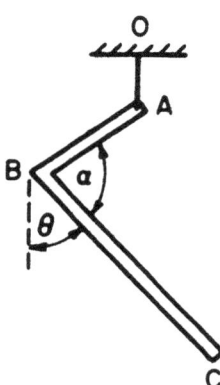

117. A uniform bar is bent to an angle-shape and suspended by means of a slender wire, as shown. If the ratio of the length of BC to BA is 5/2, determine, by application of the principle of virtual work, the angle of inclination, θ, of BC to the vertical.

Ans. $\tan \theta = \dfrac{24 \sin \alpha}{25 - 24 \cos \alpha}$

118. Employ the principle of virtual work to compute the position of equilibrium (i.e., angle θ) of the system shown in problem 33 if $W_1 > W_2$, $\alpha > \beta$.

$$\text{Ans. } \tan \theta = \frac{W_1 \sin \alpha \cos \beta - W_2 \cos \alpha \sin \beta}{(W_1 + W_2) \sin \alpha \sin \beta}$$

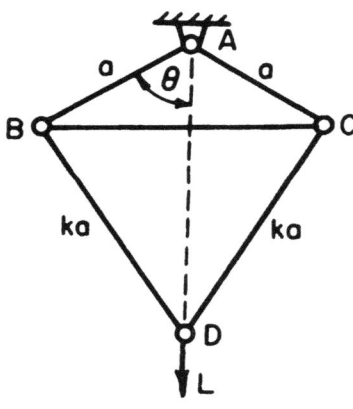

119. The latticework shown in the figure consists of bars AB and AC, each of length a, bars BD and CD, each of length ka ($k > 1$), and the horizontal cross-bar BC.

Employ the principle of virtual work to determine the compressive force, S, in BC due to the applied load L. Neglect the weight of the members in the computation.

$$\text{Ans. } S = \frac{L}{2}\left(1 + \frac{\cos \theta}{\sqrt{k^2 - \sin^2 \theta}}\right) \tan \theta$$

120. Suppose that, in the structure of problem 119, each of the chord (external) members is of weight w per unit length and the weight of BC is still negligible. Moreover, the configuration is such that $k = 3/2$ and $\theta = 60°$. What *upward* vertical force L must be exerted at D in order that member BC be unloaded?

$$\text{Ans. } L = 3.28 \, wa$$

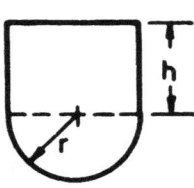

121. A solid of uniform density consists of a hemispherical base of radius r surmounted by a right circular cylinder of height h, as shown in the figure. Determine the largest ratio h/r for which the solid will be in stable equilibrium in an upright position.

$$\text{Ans. } h/r \leqslant 1/\sqrt{2}$$

122. A nonhomogeneous circular cylinder rests on a fixed horizontal plane. Show that the position is one of stable equilibrium provided the center of gravity is located vertically below the axis of the cylinder.

PROBLEMS

Solve the following by means of the potential energy approach and determine the type of equilibrium:

123. Problem 39.

124. Problem 40.

125. Problem 42.

126. Problem 44.

127. Problem 47.

128. Problem 50.

129. Problem 85.

130. Problem 117.

131. Apply the potential energy approach to determine the position of equilibrium of the system of problem 33 in which $W_1 > W_2$, $\alpha > \beta$, and the bar $S_1 S_2$ is uniform and of weight W.

Ans. $\tan \theta = \dfrac{W_1 \sin \alpha \cos \beta - W_2 \cos \alpha \sin \beta + (W/2) \sin (\alpha - \beta)}{(W_1 + W_2 + W) \sin \alpha \sin \beta}$

Chapter VIII. Equilibrium of Simple Spatial Systems

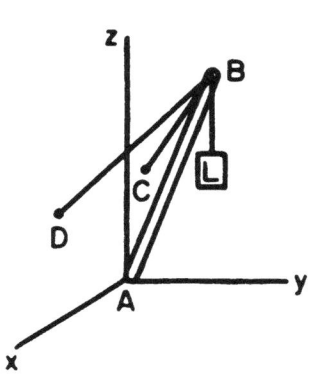

132. Bar AB is hinged at its lower end and supported by cables BC and BD. It carries a vertical load $L = 100$ lbs., as shown. The coordinates of B, C, D are (3, 4, 12), (0, 1, 5), and (2, 0, 4), respectively. Determine the forces in the cables and in the rod, assuming the weight of the latter negligible.

PROBLEMS

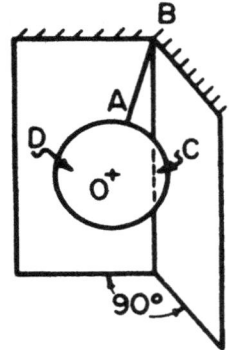

133. Sphere O, of radius a and weight W, is suspended from a corner by means of string AB, of length l. The sphere is in equilibrium with points C and D in contact with mutually perpendicular, smooth, vertical walls. Determine the forces of contact and the tension in AB.

Ans. $T_{AB} = \dfrac{W(a + l)}{\sqrt{l^2 + 2al - a^2}}$

$C = D = \dfrac{Wa}{\sqrt{l^2 + 2al - a^2}}$

134. A block of weight W rests on a plane inclined to the horizontal by angle α. The contact is rough, the coefficient of friction, μ, being sufficient to prevent sliding down the plane. Determine the magnitude P of the least horizontal force, applied to the block in the direction transverse to the incline, which will cause impending motion. Assume that the geometry of the block is such that slipping will precede tipping.

Ans. $P = W(\mu^2 \cos^2 \alpha - \sin^2 \alpha)^{1/2}$

135. Compute the forces induced in the members of the latticework shown in the figure by the three external vertical loads of magnitude P. ABC and DEF form right isosceles triangles in parallel horizontal planes a distance $3l$ apart, the right angles being at B and E, respectively. The plan view is shown in the lower diagram.

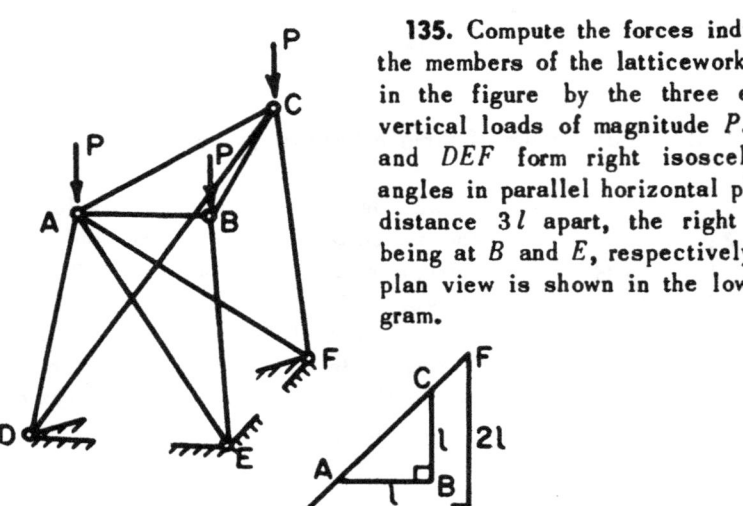

136. Three like smooth spheres, each of radius r and weight W, are placed in contact in a fixed, smooth, hemispherical bowl of radius R such that their centers lie in a horizontal plane. Find the contact forces between the spheres.

137. A fourth like sphere is added to the system of problem 135 such that it rests symmetrically on top of the first three. For what value of the radius R of the bowl will the spheres begin to separate?

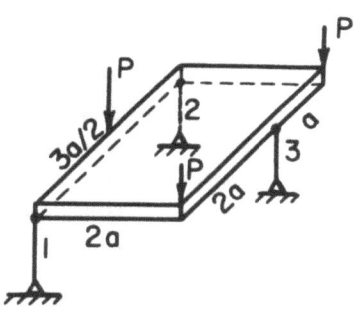

138. A uniform rectangular plate is supported in a horizontal position by three slender vertical rods and subjected to the action of three vertical forces applied as shown in the figure. Assuming the weight of the plate to be negligible in comparison to the applied loading, determine the compressive forces in the supports.

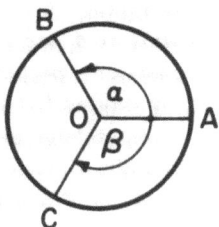

139. A uniform circular plate rests on three vertical supports, A, B, C, located at its periphery. What must be the relative position of these supports, measured by the angles α and β, as shown, in order that the compressive forces in the supports due to the weight of the plate be in the ratios $R_A : R_B : R_C = 1 : m : n$?

Ans. $\cos \alpha = \dfrac{n^2 - m^2 - 1}{2m}$

$\cos \beta = \dfrac{m^2 - n^2 - 1}{2n}$

140. A uniform plate, ABC, in the shape of an isosceles right triangle, is held in a horizontal position by means of three vertical wires attached

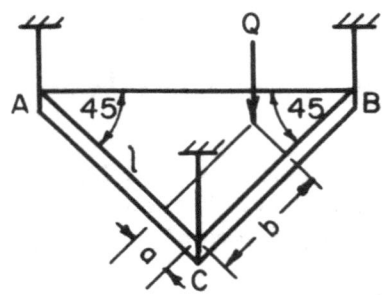

at its corners. It is subjected to a vertical load of magnitude Q applied as shown in the diagram.

Determine the tension in each of the wires if (a) the weight of the plate is neglected, (b) the weight, W, of the plate is taken into account.

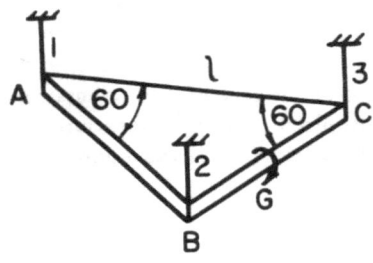

141. Find the force in each of the supporting rods of horizontal plate ABC whose shape is an equilateral triangle of side l, as shown, due to the action of a couple of magnitude G and direction such that its turning effort is about edge BC.

Ans. $F_1 = 2G/l\sqrt{3}$ (C)
$F_2 = F_3 = G/l\sqrt{3}$ (T)

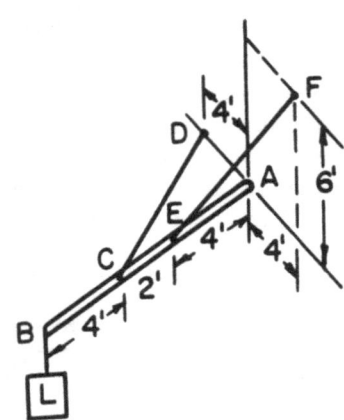

142. A uniform bar AB, 10 ft. long and weighing 50 lbs., is hinged to a wall at A and held in a horizontal position, perpendicular to the wall, by cables CD and EF, as shown. Compute the tension in the cables and the reaction at the hinge due to a load $L = 100$ lbs. suspended at end B.

143. A uniform beam, whose cross-section is a square of side a, whose length is $l = 6a$, and whose weight per unit length is $q = Q/4a$, is built

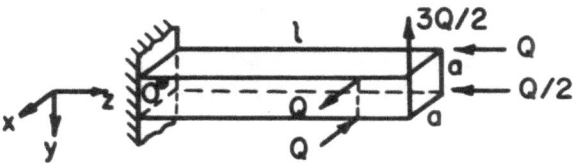

into a wall at its left end and subjected to external loading, as shown in the diagram. Determine the resultant force and moment which must be supplied by the wall at O (center of section) in order to insure static equilibrium.

Ans. $\mathbf{F} = (3Q/2)\mathbf{k}$
$\mathbf{M} = -(Qa/4)(19\mathbf{i} - 3\mathbf{j} + \mathbf{k})$

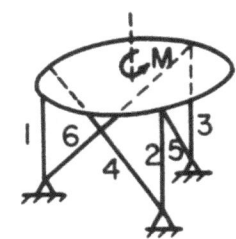

144. A circular table top of radius r is supported by three vertical bars, equally spaced along its periphery, and three inclined bars hinged to the vertical ones as shown. If each inclined bar makes an angle of $30°$ with the vertical, compute the force in each bar due to a couple of moment M about the vertical.

Ans. $T_1 = T_2 = T_3 = 2M/r\sqrt{3}$ (tension)
$T_4 = T_5 = T_6 = -4M/3r$ (compression)

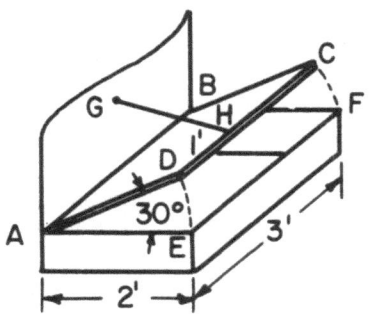

145. The lid $ABCD$ of a horizontal rectangular box is pivoted on hinges A and B and its weight of 50 lbs. is supported by chain GH, 2 ft. long, as shown in the figure. If GH is perpendicular to AB, compute the tension in the chain and the reaction at the hinges.

146. Suppose that, in problem 145, chain GH is replaced by a rigid rod FC, of negligible weight. Determine the compressive force induced in the rod.

147. A uniform rectangular slab of weight W is supported in a horizontal position by six hinged bars and subjected to external loading, as

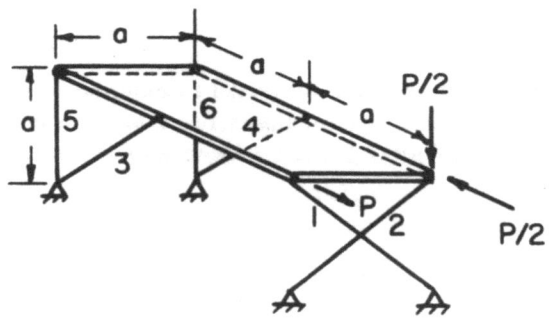

shown in the diagram. Determine the reactive force in each supporting bar.

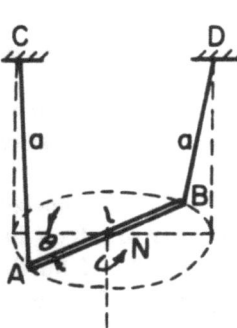

148. A uniform bar, AB, of length l and weight W, is suspended in a horizontal position by means of two vertical cables, CA and DB, each of length a. The bar is displaced from its initial position by forces statically equivalent to a couple of magnitude N about the vertical, as a result of which the bar rotates by an angle θ, as shown in the figure.

Determine the magnitude of the couple and the resulting tension in the cables corresponding to a given value of θ.

Ans. $T = Wa/2\sqrt{a^2 - l^2 \sin^2(\theta/2)}$
$N = Wl^2 \sin\theta/4\sqrt{a^2 - l^2 \sin^2(\theta/2)}$

Appendix. Center of Gravity

Determine the x and y coordinates of the centroid of each of the areas bounded by the curves:

149. The parabola $y^2 = 2px$ and the lines (a) $x = 0$, $y = b$; (b) $x = h$; (c) $y = mx$.

Ans. (a) $\bar{x} = 3b^2/20p$, $\bar{y} = 3b/4$
(b) $\bar{x} = 3h/5$, $\bar{y} = 0$
(c) $\bar{x} = 4p/5m^2$, $\bar{y} = p/m$

150. The cubical parabola $y = x^3$ and the lines (a) $x = 2$, $y = 0$; (b) $x = 0$, $y = 8$; (c) $y = 4x$ (first quadrant).

Ans. (a) $\bar{x} = 8/5$, $\bar{y} = 16/7$
(b) $\bar{x} = 4/5$, $\bar{y} = 32/7$
(c) $\bar{x} = 16/15$, $\bar{y} = 64/21$

151. The sine curve $y = \sin x$ (first arch) and the line $y = 0$.

Ans. $\bar{x} = \pi/2$, $\bar{y} = \pi/8$

152. The ellipse $x^2/a^2 + y^2/b^2 = 1$ and the coordinate axes (first quadrant).

Ans. $\bar{x} = 4a/3\pi$, $\bar{y} = 4b/3\pi$

153. The parabolas $y = x^2 - 2x - 3$ and $y = 6x - x^2 - 3$.

Ans. $\bar{x} = 2$, $\bar{y} = 1$

154. The hyperbola $x^2/a^2 - y^2/b^2 = 1$ and the lines $x = 2a$, $y = 0$ (first quadrant).

Ans. $\bar{x} \doteq 1.61a$, $\bar{y} \doteq 0.62b$

Compute the x and y coordinates of the centroid of each of the following figures:

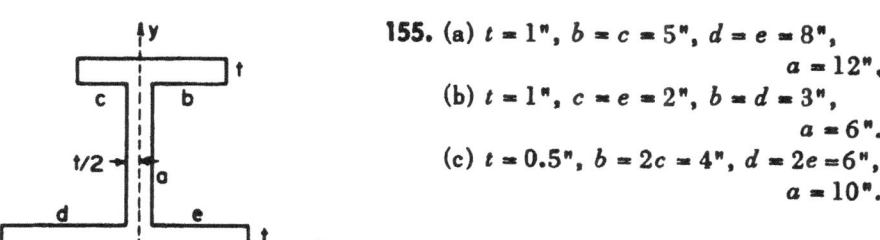

155. (a) $t = 1''$, $b = c = 5''$, $d = e = 8''$, $a = 12''$.
(b) $t = 1''$, $c = e = 2''$, $b = d = 3''$, $a = 6''$.
(c) $t = 0.5''$, $b = 2c = 4''$, $d = 2e = 6''$, $a = 10''$.

PROBLEMS

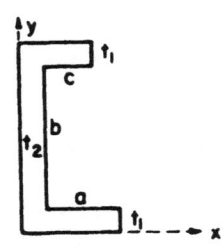

156. (a) $t_1 = t_2 = 1"$, $a = b = 2c = 6"$.
(b) $t_2 = 2t_1 = 1"$, $b = 2a = 4c = 16"$.
(c) $t_1 = 2t_2 = 2"$, $b = 2c = 3a = 12"$.

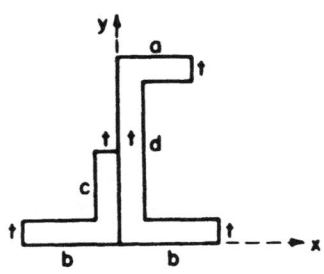

157. (a) $a = b = c = d/2 = 4t$.
(b) $t = 1"$, $d = 2b = 3a = 3c = 12"$.
(c) $t = 0.5"$, $d = 2c = 10"$, $b = 2a = 8"$.

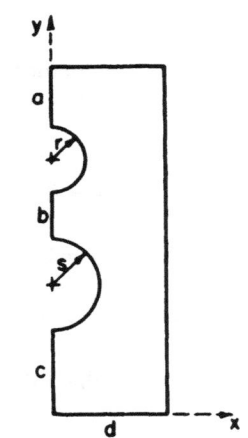

158. (a) $a = b = s = r$, $c = d = 2r$.
(b) $a = r = 1"$, $b = s = 1.5"$, $c = d = 3"$.
(c) $a = b = c = r$, $d = 2s = 3r$.

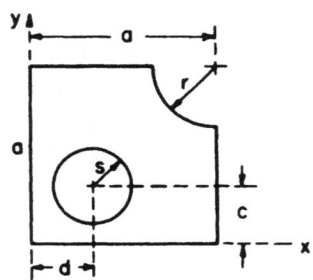

159. (a) $c = d = 2s$, $s = r$, $a = 4r$.
(b) $c = d = 2s$, $a = 2r = 8s$.
(c) $d = 3c/2 = 2s = 2"$, $r = c$, $a = 3d$.

PROBLEMS

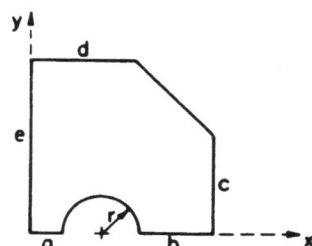

160. (a) $a = b = c = d = r$, $e = 4r$.
(b) $a = r$, $b = c = d = 2r$, $e = 3r$.
(c) $a = b = c = 2r$, $d = e = 3r$.

In each of the following problems the beam is subjected to a distributed normal load, as illustrated. Assuming the weight of the beam to be negligible, determine the magnitude of the support reactions.

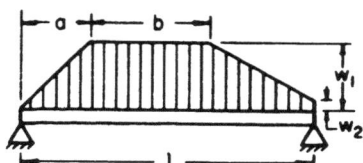

161. (a) $a = b = l/4$, $w_2 = 0$.
(b) $a = 0$, $b = l/2$, $w_1 = 2w_2$.
(c) $a = 2b = l/2$, $w_1 = 3w_2$.

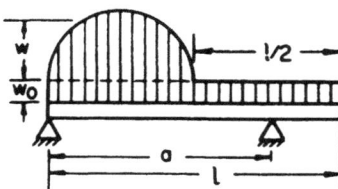

162. (a) $w = w_1 \sin 2\pi x/l$.
(b) $w = w_1 \sqrt{x(l/2 - x)}$.

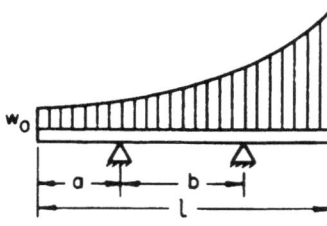

163. (a) $w = w_0 e^{kx}$.
(b) $w = w_0(1 + x^2/l^2)$.
(c) $w = w_0/(1 - x^2/3l^2)$.

PROBLEMS

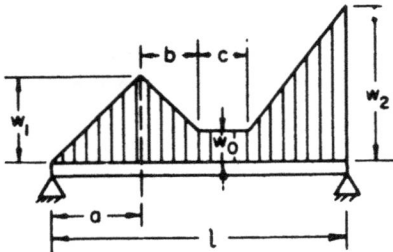

164. (a) $a = b = c = l/4$,
 $w_1 = w_2 = 2w_0$.
(b) $a = 2b = l/3$, $c = 0$,
 $w_1 = w_0 = w_2/2$.
(c) $a = 0$, $b = c = l/4$,
 $w_1 = 2w_0 = 3w_2/2$.

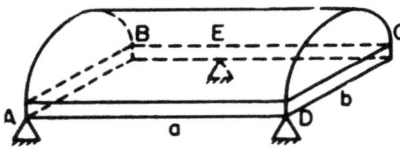

165. A horizontal rectangular plate $ABCD$ supports a semicircular solid steel cylinder, and is in turn supported at A, B, and E, as shown in the figure. If $EC = 2EB$ and the dimensions indicated are in feet, determine the support reactions. The density of steel may be taken as 0.280 lbs./in^3.

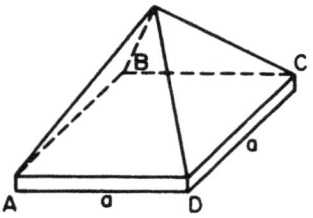

166. A square plate is heaped with sand which forms a pyramid surmounting the plate. The height of the pyramid equals the length of its base. Determine the reaction at the supports which are located at A and at the midpoint of sides BC and CD. The dimensions are in feet and the specific gravity of sand may be taken to be 1.5.

Bei Fragen zur Produktsicherheit wenden Sie sich bitte an:
If you have any questions regarding product safety,
please contact:

Walter de Gruyter GmbH
Genthiner Straße 13
10785 Berlin
productsafety@degruyterbrill.com